Contents

Chapter 1

Introduction

1.1 Commodore 64

The **Commodore 64**, also known as the **C64**, **C-64**, **C= 64**,[n 1] or occasionally **CBM 64** or **VIC-64**,[5] is an 8-bit home computer introduced in January 1982 by Commodore International. It is listed in the Guinness World Records as the highest-selling single computer model of all time,[6] with independent estimates placing the number sold between 10 and 17 million units.[7]

Volume production started in early 1982, with machines being released on to the market in August at a price of US$595 (roughly equivalent to $1,500 in 2015).[8][9] Preceded by the Commodore VIC-20 and Commodore PET, the C64 takes its name from its 64 kilobytes (65,536 bytes) of RAM, and has technologically superior sound and graphical specifications when compared to some earlier systems such as the Apple II and Atari 800, with multi-color sprites and a more advanced sound processor.

The C64 dominated the low-end computer market for most of the 1980s.[10] For a substantial period (1983–1986), the C64 had between 30% and 40% share of the US market and two million units sold per year,[11] outselling the IBM PC compatibles, Apple Inc. computers, and the Atari 8-bit family of computers. Sam Tramiel, a later Atari president and the son of Commodore's founder, said in a 1989 interview, "When I was at Commodore we were building 400,000 C64s a month for a couple of years."[12] In the UK market, the 64 faced competition from the BBC Micro and the ZX Spectrum[13] but the 64 was still one of the two most-popular computers in the UK.[14]

Part of the Commodore 64's success is because it was sold in retail stores instead of just electronics and/or computer stores. Commodore produced many of its parts in-house to control costs, including custom IC chips from MOS Technology. It has been compared to the Ford Model T automobile for its role in bringing a new technology to middle-class households via creative mass-production.[15]

Approximately 10,000 commercial software titles have been made for the Commodore 64 including development

tools, office productivity applications, and games.[16] C64 emulators allow anyone with a modern computer, or a compatible video game console, to run these programs today. The C64 is also credited with popularizing the computer demoscene and is still used today by some computer hobbyists.[17] In 2008, 17 years after it was taken off the market, research showed that brand recognition for the model was still at 87%.[6]

1.1.1 History

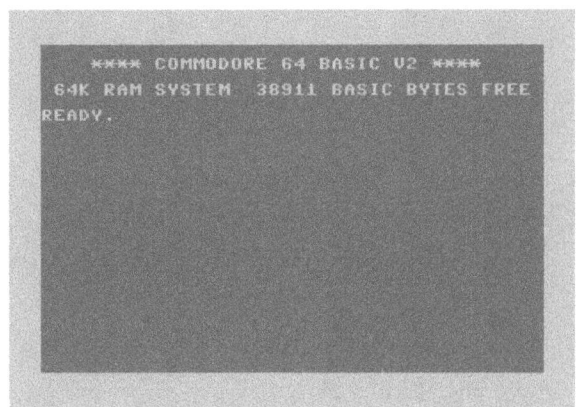

The Commodore 64 startup screen

In January 1981, MOS Technology, Inc., Commodore's integrated circuit design subsidiary, initiated a project to design the graphic and audio chips for a next generation video game console. Design work for the chips, named MOS Technology VIC-II (Video Integrated Circuit for graphics) and MOS Technology SID (Sound Interface Device for audio), was completed in November 1981.[8]

Commodore then began a game console project that would use the new chips—called the *Ultimax* or alternatively the *Commodore MAX Machine*, engineered by Yash Terakura from Commodore Japan. This project was eventually cancelled after just a few machines were manufactured for the Japanese market.

At the same time, Robert "Bob" Russell (system programmer and architect on the VIC-20) and Robert "Bob" Yannes (engineer of the SID) were critical of the current product line-up at Commodore, which was a continuation of the Commodore PET line aimed at business users. With the support of Al Charpentier (engineer of the VIC-II) and Charles Winterble (manager of MOS Technology), they proposed to Commodore CEO Jack Tramiel a true low-cost sequel to the VIC-20. Tramiel dictated that the machine should have 64 kB of random-access memory (RAM). Although 64 kB of dynamic random access memory (DRAM) cost over US$100 at the time, he knew that DRAM prices were falling, and would drop to an acceptable level before full production was reached. In November, Tramiel set a deadline for the first weekend of January, to coincide with the 1982 Consumer Electronics Show (CES).[8]

The product was code named the VIC-40 as the successor to the popular VIC-20. The team that constructed it consisted of Yash Terakura,[18] Bob Russell, Bob Yannes and David A. Ziembicki. The design, prototypes and some sample software were finished in time for the show, after the team had worked tirelessly over both Thanksgiving and Christmas weekends.

The machine incorporated Commodore BASIC 2.0 in ROM. BASIC also served as the user interface shell and was available immediately on startup at the READY prompt.

When the product was to be presented, the VIC-40 product was renamed C64.

The C64 made an impressive debut at the January 1982 Consumer Electronics Show, as recalled by Production Engineer David A. Ziembicki: "All we saw at our booth were Atari people with their mouths dropping open, saying, 'How can you do that for $595?'.".[19] The answer was vertical integration; due to Commodore's ownership of MOS Technology's semiconductor fabrication facilities, each C64 had an estimated production cost of US$135.

Market war

Upon its introduction in August 1982, the C64 faced a wide range of competing home computers.[2] With a lower price and more flexible hardware, it quickly outsold many of its competitors. In the United States the greatest competitors were the Atari 8-bit 400, the Atari 800, and the Apple II. The Atari 400 and 800 had been designed to accommodate previously stringent FCC emissions requirements and so were expensive to manufacture. The latest revision in the aging Apple II line, the Apple IIe, had higher-resolution graphics modes than the C64.[20][21] Though similar in specifications, the two computers represented differing design philosophies; as an open architecture system, upgrade capability for the Apple II was granted by internal expansion

Game cartridges for Radar Rat Race *and* International Soccer

slots, whereas the C64's comparatively closed architecture had only a single external ROM cartridge port for bus expansion. However, the Apple used its expansion slots for interfacing to common peripherals like disk drives, printers and modems; the C64 had a variety of ports integrated into its motherboard which were used for these purposes, usually leaving the cartridge port free. Commodore's was not a completely closed system, however; the company had published detailed specifications for most of their models since the PET and VIC-20 days, and the C64 was no exception.

All four machines had similar memory configurations which were standard in 1982–1983: 48K for the Apple II+[22] (upgraded within months of C64's release to 64K with the Apple IIe) and 48K for the Atari 800.[23] At upwards of $1,200,[24] the Apple II was about twice as expensive, while the Atari 800 cost $899. One key to the C64's success was Commodore's aggressive marketing tactics, and they were quick to exploit the relative price/performance divisions between its competitors with a series of television commercials after the C64's launch in late 1982.[25] The company also published detailed documentation to help developers,[26] while Atari initially kept technical information secret.[27] At a mid-1984 conference of game developers and experts at Origins Game Fair, Dan Bunten, Sid Meier ("the computer of choice right now"), and a representative of Avalon Hill all stated that they were developing games for the 64 first as the most promising market.[28] In April 1986 *Computer Gaming World* published a survey of ten game publishers which found that they planned to release forty-three Commodore 64 games that year, compared to nineteen for Atari and forty-eight for Apple II,[29] and that year Alan Miller stated that Accolade developed first for the C64 because "it will sell the most on that system".[30]

Commodore sold the C64 not only through its network of authorized dealers, but also through department stores, discount stores, toy stores and college bookstores. The C64 had a built-in RF modulator and thus could be plugged into any television set. This allowed it (like its predecessor, the VIC-20) to compete directly against video game consoles such as the Atari 2600. Like the Apple IIe, the C64 can also output a composite video signal (avoiding the RF modulator) that can be plugged into a specialized monitor for a sharper picture. Unlike the IIe, the C64's NTSC output capability also includes separate luminance/chroma signal output equivalent to (and electrically compatible with) S-Video, for connection to the Commodore 1702 monitor, providing even better video quality than a composite signal.

Aggressive pricing of the C64 is considered to have been a major catalyst in the North American video game crash of 1983. In January 1983, Commodore offered a $100 rebate in the United States on the purchase of a C64 to anyone trading in another video game console or computer.[31] To take advantage of this rebate, some mail-order dealers and retailers offered a Timex Sinclair 1000 for as little as $10 with purchase of a C64, so the consumer could send the TS1000 to Commodore, collect the rebate, and pocket the difference; Timex Corporation departed the computer market within a year. Commodore's tactics soon led to a price war with the major home computer manufacturers. The success of the VIC-20 and C64 contributed significantly to the exit of Texas Instruments and other smaller competitors from the field.

The price war with Texas Instruments was seen as a personal battle for Commodore president Jack Tramiel.[32] In June 1983 Commodore lowered the C64's list price to $300, and some stores sold the computer for $199. At one point, the company was selling as many C64s as all computers sold by the rest of the industry combined, while TI lost money by selling the 99/4A for $99.[33] TI's subsequent demise in the home computer industry in October 1983 was seen as revenge for TI's tactics in the electronic calculator market in the mid-1970s, when Commodore was almost bankrupted by TI.[34] *Computer Gaming World* stated in January 1985 that companies such as Epyx that survived the video game crash did so because they "jumped on the Commodore bandwagon early".[35]

In Europe, the primary competitors to the C64 were British-built computers: the Sinclair ZX Spectrum, the BBC Micro and the Amstrad CPC464. In the UK, the 48K Spectrum had not only been released a few months ahead of the C64's early 1983 debut, but it was also selling for £175, less than half the C64's £399 price. The Spectrum quickly became the market leader and Commodore had an uphill struggle against it. The C64 did however go on to rival the Spectrum in popularity in the latter half of the 1980s. Adjusted to the size of population, the popularity of Commodore 64 was the highest in Finland where it was subsequently marketed as "the computer of the republic".[36]

Although rumors spread in late 1983 that Commodore would discontinue the C64,[37] Commodore sold about one million C64s in 1985 and a total of 3.5 million by mid-1986. Although the company reportedly attempted to discontinue the C64 more than once in favor of more expensive computers such as the Commodore 128, demand remained strong.[38][39] In 1986 Commodore introduced the 64c,[40] a redesigned 64, which *Compute!* saw as evidence that—contrary to C64 owners' fears that the company would abandon them in favor of the Amiga and 128—"the 64 refuses to die".[41] Its introduction also meant that Commodore raised the price of the C64 for the first time, which the magazine cited as the end of the home-computer price war.[42] Software sales also remained strong; MicroProse, for example, in 1987 cited the Commodore and IBM PC markets as its top priorities.[43]

By 1988, Commodore was still selling between one and one and a half million C64s worldwide every year,[44] although Epyx CEO David Shannon Morse cautioned that "there are no new 64 buyers, or very few. It's a consistent group that's not growing ... it's going to shrink as part of our business".[45] One computer-gaming executive stated that the Nintendo Entertainment System's enormous popularity—seven million sold that year, almost as many as the number of C64s sold in its first five years—had stopped the C64's growth, and Trip Hawkins stated that Nintendo was "the last hurrah of the 8-bit world".[46] Although in the United States demand for the C64 had dropped off by 1990, it continued to be popular in the UK and other European countries. In the end it was not lack of demand or the cost of the C64 itself (still profitable at a retail price point between £44 and £50), but the cost of producing the drive that ended the machine's long run. In March 1994, at CeBIT in Hanover, Germany, Commodore announced that the C64 would be finally discontinued in 1995, [47] noting that the Commodore 1541 cost more than the C64 itself.[47] However, only one month later, in April 1994, the company filed for bankruptcy.

The C64 family

Commodore MAX Machine

In 1982 Commodore released the Commodore MAX Machine in Japan. It was called the Ultimax in the United States, and VC-10 in Germany. The MAX was intended to be a game console with limited computing capability, and was based on a very cut-down version of the hardware family later used in the C64. The MAX was discontinued months after its introduction because of poor sales in Japan.[48]

1983 saw Commodore attempt to compete with the Apple II's hold on the U.S. education market with the Educator 64,[49] essentially a C64 and "greenscale" monochrome monitor in a PET case. Schools preferred the all-in-one metal construction of the PET over the standard C64's separate components, which could be easily damaged, vandalized or stolen.[50] Schools did not prefer the Educator 64 to the wide range of software and hardware options the Apple IIe was able to offer, and it was produced in limited quantities.[51]

Commodore SX-64

Also in 1983, Commodore released the SX-64, a portable version of the C64. The SX-64 has the distinction of being the first *full-color* portable computer. While earlier computers using this form factor only incorporated monochrome "green screen" displays, the base SX-64 unit featured a 5 in (130 mm) color cathode ray tube (CRT) and an integrated 1541 floppy disk drive. Unlike most other C64s, the SX-64 did not have a cassette connector.[52]

In 1984, Commodore released the Commodore Plus/4. It has a higher-color display, a newer implementation of Commodore BASIC (V3.5), and built-in software in what was positioned as an inexpensive business oriented system. However, it is incompatible with the C64, and the burgeoning influence of the IBM PC on the personal computer market market rendered the limited business software of the

Plus/4 system of marginal value. The Plus/4 lacks hardware sprite capability and lacks a SID chip, thus underperforming in two of the areas that had made the C64 successful.

Two designers at Commodore, Fred Bowen and Bil Herd, were determined to rectify the problems of the Plus/4. They intended that the eventual successors to the C64— the Commodore 128 and 128D computers (1985)—were to build upon the C64, avoiding the Plus/4's flaws.[53][54] The successors had many improvements such as a structured BASIC with graphics and sound commands, 80-column display ability, and full CP/M compatibility. The decision to make the Commodore 128 plug compatible with the C64 was made quietly by Bowen and Herd, software and hardware designers respectively, without the knowledge or approval by the management in the post Jack Tramiel era. The designers were careful not to reveal their decision until the project was too far along to be challenged or changed and still make the impending Consumer Electronics Show (CES) show in Las Vegas.[53] Upon learning that the C128 was designed to be compatible with the C64, Commodore's marketing department independently announced that the C128 would be 100% compatible with the C64, thereby raising the bar for C64 support. In a case of malicious compliance, the 128 design was altered to include a separate "64 mode" using a complete C64 environment to ensure total compatibility.

Commodore 64c with 1541-II floppy disk drive and 1084S monitor displaying television-compatible S-video

In 1986, Commodore released the 64c computer, which is functionally identical to the original. The exterior design was remodeled in the sleeker style of the Commodore 128.[39] The 64c's modifications are more than skin-deep with new versions of the SID, VIC and I/O chips being deployed—with the core voltage reduced from 12V to 9V. In the United States, the 64c was often bundled with the third-party GEOS graphical user interface (GUI) based operating system, as well as the software needed to access QuantumLink. The Commodore 1541 disk drive received a matching face-lift resulting in the 1541c. Later a smaller, sleeker 1541-II model was introduced along with the 800 kB 3.5-inch microfloppy 1581.

Commodore 64 Games System "C64GS"

In 1990, the C64 was repackaged in the form of a game console, called the C64 Games System (C64GS), with most external connectivity removed.[55] A simple modification to the 64c's motherboard was made to allow cartridges to be inserted from above. A modified ROM replaced the BASIC interpreter with a boot screen to inform the user to insert a cartridge. Designed to compete with the Nintendo Entertainment System and the Sega Master System, it suffered from very low sales compared to its rivals. It was another commercial failure for Commodore, and it was never released outside Europe.

In 1990, an advanced successor to the C64, the Commodore 65 (also known as the "C64DX"), was prototyped, but the project was canceled by Commodore's chairman Irving Gould in 1991. The C65's specifications are very good for an 8-bit computer, bringing specs comparable to the 16-bit Apple IIgs. For example, it can display 256 colors on screen, while OCS based Amigas can only display 64 in HalfBrite mode (32 colors and half-bright transformations). Although no specific reason was given for the C65's cancellation, it would have competed in the marketplace with Commodore's lower end Amigas and the Commodore CDTV.

C64 clone

In the middle of 2004, after an absence from the marketplace of more than 10 years, PC manufacturer Tulip Computers BV (owners of the Commodore brand since 1997) announced the C64 Direct-to-TV (C64DTV), a joystick-based TV game based on the C64 with 30 games built into ROM. Designed by Jeri Ellsworth, a self-taught computer designer who had earlier designed the modern C-One C64 implementation, the C64DTV was similar in concept to other mini-consoles based on the Atari 2600 and Intellivision which had gained modest success earlier in the decade. The product was advertised on QVC in the United

States for the 2004 holiday season.[56] By *"hacking"* the circuit board, it is possible to attach C1541 floppy disk drives, hard drives, second joysticks, and PS/2-keyboards to these units, which gives the DTV devices nearly all of the capabilities of a full Commodore 64. The DTV hardware is also used in the mini-console *Hummer*, sold at RadioShack in mid-2005.

Newer compatible hardware

C64 enthusiasts still develop new hardware, including Ethernet cards,[57] specially adapted hard disks and flash card interfaces (sd2iec).[58]

Brand reuse

The C64 "Web.it" Internet Computer

In 1998, the C64 brand was reused for the "Web.it Internet Computer",[59][60] a low-powered (even for the time) Internet-oriented, all-in-one x86 PC running Windows 3.1. Despite its "Commodore 64" nameplate, the "C64 Web.it" is not directly compatible with the original (except via included emulation software), nor does it share its appearance.

PC clones branded as C64x sold by Commodore USA, LLC, a company licensing the Commodore trademark,[61][62] began shipping in June 2011.[63][64] The C64x has a case resembling the original C64 computer, but- as with the "Web.it"- it is based on x86 architecture and is not compatible with the Commodore 64 on either hardware or software levels.

Virtual Console

Several Commodore 64 games were released on the Nintendo Wii's Virtual Console service in Europe and North America only. The games were removed from the service as of August 2013 for unknown reasons.

1.1.2 Software

Main article: Commodore 64 software

In 1982, the C64's graphics and sound capabilities were rivaled only by the Atari 8-bit family, and appeared exceptional when compared with the widely publicised Atari VCS and Apple II.

The C64 is often credited with starting the computer subculture known as the demoscene (see Commodore 64 demos). It is still being actively used in the demoscene,[65] especially for music (its sound chip even being used in special sound cards for PCs, and the Elektron SidStation synthesizer).

Even though other computers quickly caught up with it, the C64 remained a strong competitor to the later video game consoles Nintendo Entertainment System (NES) and Sega Master System, thanks in part to its by-then established software base, especially outside North America, where it comprehensively outsold the NES.

BASIC

Main article: Commodore BASIC
As is common for home computers of the early 1980s,

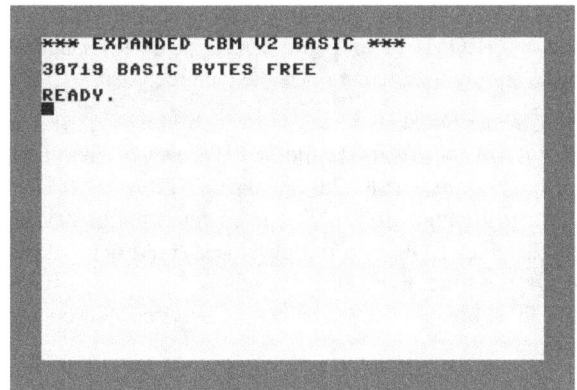

The Simons' BASIC start-up screen. Note the altered background and text colors (vs the ordinary C64 blue tones), and the 8 kB reduction of available BASIC program memory due to the address space used by the cartridge.

the C64 incorporates a ROM-based version of the BASIC programming language. There is no operating system as such. The KERNAL is accessed via BASIC commands. The disk drive has its own microprocessor, much like the earlier CBM/PET systems and the Atari 400 & Atari 800. This means that no memory space is dedicated to running a disk operating system, as was the case with earlier systems such as the Apple II and TRS-80.

Commodore BASIC 2.0 is used instead of the more advanced BASIC 4.0 from the PET series, since C64 users were not expected to need the disk-oriented enhancements of BASIC 4.0. The company did not expect many to buy a disk drive, and using BASIC 2.0 simplified VIC-20 owners' transition to the 64.[66] "The choice of BASIC 2.0 instead of 4.0 was made with some soul-searching, not just at random. The typical user of a C64 is not expected to need the direct disk commands as much as other extensions and the amount of memory to be committed to BASIC were to be limited. We chose to leave expansion space for color and sound extensions instead of the disk features. As a result, you will have to handle the disk in the more cumbersome manner of the 'old days'."[67]

The version of BASIC is limited and does not include specific commands for sound or graphics manipulation, instead requiring users to use the "PEEK and POKE" commands to access the graphics and sound chip registers directly. To provide extended commands, including graphics and sound, Commodore produced two different cartridge-based extension to BASIC 2.0: Simons' BASIC and Super Expander 64.

Other languages available for the C64 include Pascal, C, [68][69] Logo, Forth, and FORTRAN. Compilers for BASIC 2.0 such as Petspeed 2 (from Commodore), Blitz (from Jason Ranheim) and Turbo Lightning (from Ocean Software) were produced. While much of the first generation of C64 software used the standard BASIC, after 1983 almost all professionally produced programs were written in assembly language, either cross developed on a larger computer, or directly on the C64 using a machine code monitor or an assembler. This maximized speed and minimized memory use.

Alternative operating systems

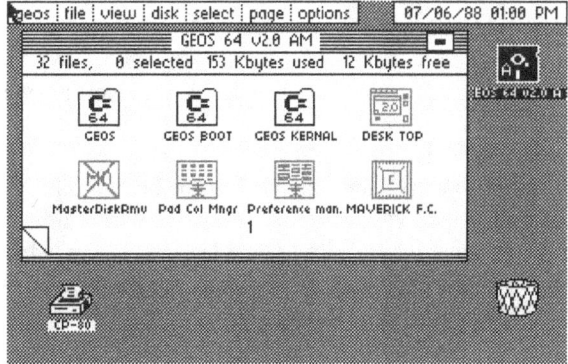

GEOS for the Commodore 64

Many third party operating systems have been developed for the C64. As well as the original GEOS, two third-party

GEOS-compatible systems have been written: Wheels and GEOS megapatch. Both of these require hardware upgrades to the original C64. Several other operating systems are or have been available, including WiNGS OS, the Unix-like LUnix, operated from a command-line, and the embedded systems OS Contiki, with full GUI. Other less well known OSes include ACE, Asterix, DOS/65 and GeckOS.

A version of CP/M was released, but this requires the addition of an external Z80 processor to the expansion bus. Furthermore, the Z80 processor is underclocked to be compatible with the C64's memory bus, so performance is poor compared to other CP/M implementations. C64 CP/M and C128 CP/M both suffer a lack of software; although most commercial CP/M software can run on these systems, software media is incompatible between platforms. The low usage of CP/M on Commodores means that software houses saw no need to invest in mastering versions for the Commodore disk format.

Networking software

During the 1980s, the Commodore 64 was used to run bulletin board systems using software packages such as Bizarre 64, Blue Board, C-Net, Color 64, CMBBS, C-Base, DMBBS, Image BBS, and The Deadlock Deluxe BBS Construction Kit, often with sysop-made modifications. These boards sometimes were used to distribute cracked software. As late as December 2013, there were 25 such Bulletin Board Systems in operation, reachable via the Telnet protocol.[70]

There were major commercial online services, such as Compunet (UK), CompuServe (US – later bought by America Online), The Source (US) and Minitel (France) among many others. These services usually required custom software which was often bundled with a modem and included free online time as they were billed by the minute.

Quantum Link (or Q-Link) was a U.S. and Canadian online service for Commodore 64 and 128 personal computers that operated from November 5, 1985, to November 1, 1994. It was operated by Quantum Computer Services of Vienna, Virginia, which in October 1991 changed its name to America Online, and continued to operate its AOL service for the IBM PC compatible and Apple Macintosh. Q-Link was a modified version of the PlayNET system, which Control Video Corporation (CVC, later renamed Quantum Computer Services) licensed.

Online gaming

Main article: History of massively multiplayer online games

The first graphical character-based interactive environment is *Club Caribe*. First released as *Habitat* in 1988, *Club Caribe* was introduced by LucasArts for Q-Link customers on their Commodore 64 computers. Users could interact with one another, chat and exchange items. Although the game's open world was very basic, its use of online avatars (already well-established off-line by *Ultima* and other games) and combination of chat and graphics was revolutionary. Online graphics in the late 1980s were severely restricted by the need to support modem data transfer rates as slow as 300 bits per second. Habitat's graphics were stored locally on floppy disk, eliminating the need for network transfer.[71]

1.1.3 Hardware

CPU and memory

Main article: MOS Technology 6510

The C64 uses an 8-bit MOS Technology 6510 microprocessor. This is a close derivative of the 6502 with an added 6-bit internal I/O port that in the C64 is used for two purposes: to bank-switch the machine's read-only memory (ROM) in and out of the processor's address space, and to operate the datasette tape recorder.

The C64 has 64½ kB of RAM, of which 1 kb of ½ bytes are color RAM for text mode and 38 kB are available to built-in Commodore BASIC 2.0 on startup. There is 20 kB of ROM, made up of the BASIC interpreter, the kernel, and the character ROM. As the processor could only address 64 kB at a time, the ROM was mapped into memory and only 38,911 bytes of RAM (plus 4 kb between ROMs) were available at startup.

If a program does not use the BASIC interpreter, RAM can be read as well as written over that ROM's location. However, this means the character ROM is not available, and the RAM in its place is instead used for the character glyphs. Normally, this RAM is uninitialized, which then results in nothing but random patterns appearing on the screen. This is solved by copying the character ROM into RAM. This had two benefits: the standard typeface can be rewritten, and character codes can be rewritten as picture elements.

Most C64 games have been written in this way, using low resolution, which requires much less processor time and memory. Furthermore, picture elements can be reused, saving even more memory. The same technique is used on the NES.

Graphics

Main article: MOS Technology VIC-II

The graphics chip, VIC-II, features 16 colors, eight hardware sprites per scanline (enabling up to 112 sprites per PAL screen), scrolling capabilities, and two bitmap graphics modes. The standard text mode features 40 columns, like most Commodore PET models; the built in character encoding is not standard ASCII but PETSCII, an extended form of ASCII-1963.

Most screenshots show borders around the screen, which is a feature of the VIC-II chip. By utilizing interrupts to reset various hardware registers on precise timings it was possible to place graphics within the borders and thus utilize the full screen.[72]

There are two low-resolution and two bitmapped modes. Multicolor bitmapped mode has an addressable screen of 160×200 pixels, with a maximum of four colors per 4×8 character block. High-resolution bitmapped mode has an addressable screen of 320×200 pixels, with a maximum of two colors per 8×8 character block.

Multicolor low-resolution has a screen of 160×200 pixels, 40×25 addressable with four colors per 8×8 character block; high resolution "low resolution" has a screen of 320×200 pixels, 40×25 addressable with two colors per 8×8 character block. Most C64 video games are multicolor low-resolution; this allows only block-by-block character animation due to the limited addressable space. However, further innovation allows video chips to automate sprites and vertical and horizontal scrolling pixel-by-pixel, allowing graphics to work smoothly and quickly regardless of the video mode. Some animation, like bullets, use character animation when sprites are unavailable.

Sound

Main article: MOS Technology SID

The SID chip has three channels, each with its own ADSR envelope generator and filter capabilities. ring modulation makes use of channel N°3, to work with the other two channels. Bob Yannes developed the SID chip and later co-founded synthesizer company Ensoniq. Yannes criticized other contemporary computer sound chips as "primitive, obviously...designed by people who knew nothing about music". Often the game music has become a hit of its own among C64 users. Well-known composers and programmers of game music on the C64 are Rob Hubbard, Tel Jeroen, David Whittaker, Chris Hülsbeck, Ben Daglish, Martin Galway and David Dunn among many others. Due

to the chip's three channels, chords are played as arpeggios, coining the C64's characteristic lively sound. It was also possible to continuously update the master volume with sampled data to enable the playback of 4-bit digitized audio. As of 2008, it became possible to play four channel 8-bit audio samples, 2 SID channels and still use filtering.[73]

There are two versions of the SID chip: the 6581 and the 8580. The MOS Technology 6581 was used in the original ("breadbox") C64s, the early versions of the 64c, and the Commodore 128. The 6581 was replaced with the MOS Technology 8580 in 1987. The 6581 sound quality is a little crisper, and many Commodore 64 fans have said that they prefer its sound. The main difference between the 6581 and the 8580 is the supply voltage. The 6581 uses a 12 volt supply—the 8580, a 9 volt supply. A modification can be made to use the 6581 in a newer 64c board (which uses the 9 volt chip).

The SID chip's distinctive sound has allowed it to retain a following long after its host computer was discontinued. A number of audio enthusiasts and companies have designed SID-based products as add-ons for the C64, x86 PCs, and standalone or MIDI music devices such as the Elektron Sid-Station. These devices use chips taken from excess stock, or removed from used computers.

In 2007, Timbaland's extensive use of the SidStation led to the plagiarism controversy for "Block Party" and "Do It" (written for Nelly Furtado).

Hardware revisions

Cost reduction was the driving force behind the C64's vintage motherboard revisions. Reducing manufacturing costs was vitally important to Commodore's survival during the price war and leaner years of the 16-bit era. The C64's original (NMOS based) motherboard would go through two major redesigns, (and numerous sub-revisions) exchanging positions of the VIC-II, SID and PLA chips. Initially, a large portion of the cost was eliminated by reducing the number of discrete components, such as diodes and resistors, which enabled the use of a smaller printed circuit board.

The case is made from ABS plastic which may become brown with time. This can be reversed by using the public domain chemical mix "Retr0bright".

ICs The VIC-II was manufactured with 5 micrometer NMOS technology[8] and was clocked at either 17.73447 MHz (PAL) or 14.31818 MHz (NTSC). Internally, the clock was divided down to generate the dot clock (about 8 MHz) and the two-phase system clocks (about 1 MHz; the exact pixel and system clock speeds are slightly different between NTSC and PAL machines). At such high clock rates,

An early C64 motherboard (Rev A PAL 1982).

A C64C motherboard ("C64E" Rev B PAL 1992).

the chip generated a lot of heat, forcing MOS Technology to use a ceramic dual in-line package called a "CERDIP". The ceramic package was more expensive, but it dissipated heat more effectively than plastic.

After a redesign in 1983, the VIC-II was encased in a plastic dual in-line package, which reduced costs substantially, but it did not totally eliminate the heat problem.[8] Without a ceramic package, the VIC-II required the use of a heat sink. To avoid extra cost, the metal RF shielding doubled as the heat sink for the VIC, although not all units shipped with this type of shielding. Most C64s in Europe shipped with a cardboard RF shield, coated with a layer of metal foil. The effectiveness of the cardboard was highly questionable, and worse still it acted as an insulator, blocking airflow which trapped heat generated by the SID, VIC, and PLA chips.

The SID was originally manufactured using NMOS at 7 and in some areas 6 micrometers.[8] The prototype SID and some very early production models featured a ceramic dual in-line package, but unlike the VIC-II, these are extremely rare as the SID was encased in plastic when production started in early 1982.

Motherboard   In 1986, Commodore released the last revision to the classic C64 motherboard. It was otherwise identical to the 1984 design, except for the two 64 kilobit × 4 bit DRAM chips that replaced the original eight 64 kilobit × 1 bit ICs.

After the release of the Commodore 64c,[74] MOS Technology began to reconfigure the original C64's chipset to use HMOS production technology. The main benefit of using HMOS was that it required less voltage to drive the IC,

which consequently generates less heat. This enhanced the overall reliability of the SID and VIC-II. The new chipset was renumbered to 85xx to reflect the change to HMOS.

In 1987, Commodore released a 64c variant with a highly redesigned motherboard commonly known as a "short board". The new board used the new HMOS chipset, featuring a new 64-pin PLA chip. The new "SuperPLA", as it was dubbed, integrated many discrete components and transistor–transistor logic (TTL) chips. In the last revision of the 64c motherboard, the 2114 color RAM was integrated into the SuperPLA.

Power supply

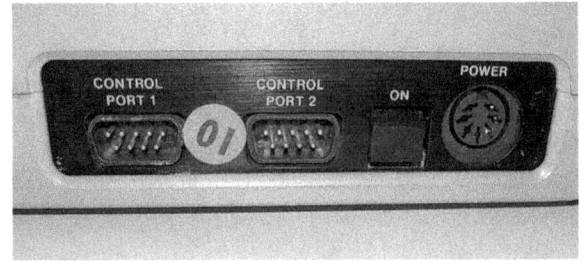

Joystick ports, power switch, power inlet

The C64 used an external power supply, a conventional transformer with multiple tappings (as opposed to switch mode, the type now used on PC power supplies), encased in an epoxy resin gel which discouraged tampering but tended to increase the heat level during use.

This saved space within the computer's case and allowed international versions to be more easily manufactured. The 1541-II and 1581 disk drives, along with various third-party clones, also come with their own external power supply "bricks", as did most peripherals leading to a "spaghetti" of cables and the use of numerous double adapters by users. These power supplies are notorious for failing over time; the computer reportedly had a 30% return rate in late 1983, compared to the 5-7% the industry considered acceptable.[75][76]

Commodore later changed the design, omitting the gel. The follow-on model, the Commodore 128, used a larger, improved power supply that included a fuse. The power supply that came with the Commodore REU was similar to that of the Commodore 128's unit, providing an upgrade for customers who purchased that accessory.

Specifications

Internal hardware

- Microprocessor CPU:

- MOS Technology 6510/8500 (the 6510/8500 is a modified 6502 with an integrated 6-bit I/O port)
- Clock speed: 0.985 MHz (PAL) or 1.023 MHz (NTSC)

- Video: MOS Technology VIC-II 6567/8562 (NTSC), 6569/8565 (PAL)

  - 16 colors
  - Text mode: 40×25 characters; 256 user-defined chars (8×8 pixels, or 4×8 in multicolor mode); or extended background color; 64 user-defined chars with 4 background colors, 4-bit color RAM defines foreground color
  - Bitmap modes: 320×200 (2 unique colors in each 8×8 pixel block),[77] 160×200 (3 unique colors + 1 common color in each 4×8 block)[78]
  - 8 hardware sprites of 24×21 pixels (12×21 in multicolor mode)
  - Smooth scrolling, raster interrupts

- Sound: MOS Technology 6581/8580 SID

  - 3-channel synthesizer with programmable ADSR envelope
  - 8 octaves
  - 4 waveforms per audio channel: triangle, sawtooth, variable pulse, noise
  - Oscillator synchronization, ring modulation
  - Programmable filter: high pass, low pass, band pass, notch filter

- Input/Output: Two 6526 Complex Interface Adapters

  - 16 bit parallel I/O
  - 8 bit serial I/O
  - 24-hours (AM/PM) Time of Day clock (TOD), with programmable alarm clock[79]
  - 16 bit interval timers

- RAM:

  - 64 kB, of which 38 kB (minus 1 byte) were available for BASIC programs
  - 512 bytes color RAM (memory allocated for screen color data storage) [80]
  - Expandable to 320 kB with Commodore 1764 256 kB RAM Expansion Unit (REU); although only 64 kB directly accessible; REU mostly intended for GEOS. REUs of 128 kB and 512 kB, originally designed for the C128, were also

available, but required the user to buy a stronger power supply from some third party supplier; with the 1764 this was included. Creative Micro Designs also produced a 2 MB REU for the C64 and C128, called the 1750 XL. The technology actually supported up to 16 MB, but 2 MB was the biggest one officially made. Expansions of up to 16 MB were also possible via the CMD SuperCPU.

- ROM:

  - 20 kB (9 kB Commodore BASIC 2.0; 7 kB KERNAL; 4 kB character generator, providing two 2 kB character sets)

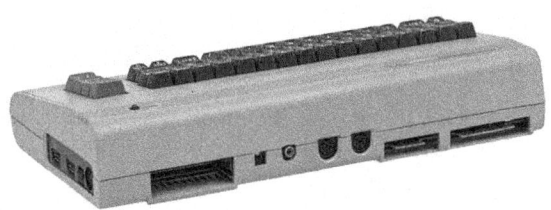

Commodore 64 ports (from left: Joy1, Joy2, Power, ROM cartridge, RF-adj, RF, A/V, 488, Tape, User)

I/O ports and power supply

- I/O ports:[81]

  - ROM cartridge expansion slot (44-pin slot for edge connector with 6510 CPU address/data bus lines and control signals, as well as GND and voltage pins;[82] used for program modules and memory expansions, among others)
  - Integrated RF modulator antenna output via a RCA connector. The used channel could be adjusted from number 36 with the potentiometer to the left.
  - 8-pin DIN connector containing composite video output, separate Y/C outputs and sound input/output. Beware that this is the 262° (horseshoe) version of the plug, not the 270° circular version. Some early C64 units use a 5-pin DIN connector that carries composite video and luminance signals, but lacks a chroma signal.[83]
  - Serial bus (serial version of IEEE-488, 6-pin DIN plug) for CBM printers and disk drives
  - PET-type Commodore Datassette 300 baud tape interface (edge connector with digital cassette motor/read/write/key-sense signals, Ground and +5V DC lines. The cassette motor is controlled

by a +5V DC signal from the 6502 CPU. The 9V AC input is transformed into unregulated 6.36V DC[84] which is used to actually power the cassette motor.[85]

- User port (edge connector with TTL-level signals, for modems and so on.; byte-parallel signals which can be used to drive third-party parallel printers, among other things, 17 logic signals, 7 Ground and voltage pins, including 9V AC)

- 2 × screwless DE9M game controller ports (compatible with Atari 2600 controllers), each supporting five digital inputs and two analog inputs. Available peripherals included digital joysticks, analog paddles, a light pen, the Commodore 1351 mouse, and graphics tablets such as the KoalaPad.

- Power supply:

  - 5V DC and 9V AC from an external "power brick", attached to a 7-pin female DIN-connector on the computer.[86]

The 9 volt AC is used to supply power via a charge pump to the SID sound generator chip, provide 6.8V via a rectifier to the cassette motor, a "0" pulse for every positive half wave to the time-of-day (TOD) input on the CIA chips, and 9 volts AC directly to the user-port. Thus, as a minimum, a 12 V square wave is required. But a 9 V sine wave is preferred.[87][88]

Memory map   Note that even if I/O chips like VIC-II only uses 64 positions in the memory address space, it will occupy 1,024 addresses because some address bits are left undecoded.[89]

Peripherals   See also: Commodore 64 peripherals

- Commodore 1541 Floppy Drive

- Commodore 1541C Floppy Drive

- Commodore 1541-II Floppy Drive

- Commodore 1530 Datasette

- Commodore MPS-802 Dot matrix printer

- Commodore VIC-Modem

- Commodore 1351 Mouse

- Commodore 1702 video monitor

1.1.4   Reception

BYTE in July 1983 stated that "the 64 retails for $595. At that price it promises to be one of the hottest contenders in the under-$1000 personal computer market". It described SID as "a true music synthesizer...the quality of the sound has to be heard to be believed", while criticizing the use of Commodore BASIC 2.0, the floppy disk performance which is "even slower than the Atari 810 drive", and Commodore's quality control.[90]

1.1.5   Emulators

Software Commodore 64 emulators include the open source VICE, and CCS64.

1.1.6   See also

- C-One – Programmable FPGA clone

- C64 Direct-to-TV – ASIC clone with 30 built in games

- Commodore 64 demos

- Commodore 64 peripherals

- Commodore 128 – the Commodore 64's descendant

- Commodore MAX Machine – the Commodore 64's predecessor

- Commodore SX-64 – Portable version

- History of personal computers

- IDE64 – P-ATA interface cartridge for the C64

- List of Commodore 64 games

- SuperCPU - CPU upgrade for C64 and C128

1.1.7   Notes

[1] The "C=" represents the graphical part of the logo.

1.1.8   References

[1] World of Commodore Brochure (1983)

[2] July 1982 Commodore brochure

[3] "How many Commodore 64 computers were sold?". Retrieved February 1, 2011.

[4] Reimer, Jeremy. "Personal Computer Market Share: 1975–2004". Retrieved July 17, 2009.

[5] VIC 64 Användarmanual. Image of Swedish edition of the VIC 64 user's manual. Retrieved on March 12, 2007.

[6] "The Commodore 64, that '80s computer icon, lives again". Retrieved November 17, 2014.

[7] "How many Commodore 64 computers were really sold?".

[8] Perry, Tekla S.; Wallich, Paul (March 1985). "Design case history: the Commodore 64" (PDF). *IEEE Spectrum*: 48–58. ISSN 0018-9235. Retrieved November 12, 2011.

[9] "IEEE Spectrum". March 1985. Retrieved November 3, 2014.

[10] "Inside the Commodore 64". *PCWorld*. November 4, 2008. Retrieved November 17, 2014.

[11] Reimer, Jeremy. "Total share: 30 years of personal computer market share figures". Ars Technica. Retrieved October 10, 2014.

[12] Naman, Mard (September 1989). "From Atari's Oval Office An Exclusive Interview With Atari President Sam Tramiel". *STart* (San Francisco: Antic Publishing) **4** (2): p. 16.

[13] "Commodore 64 Turns 30".

[14] Zuckerman, Faye (17 Nov 1984). "Now Playing". *BillBoard* (Billboard Publications) **96** (46): p. 23. ISSN 0006-2510. Retrieved June 8, 2015.

[15] Kahney, Leander (September 9, 2003). "Grandiose Price for a Modest PC". CondéNet, Inc. Retrieved September 13, 2008.

[16] "Impact of the Commodore 64: A 25th Anniversary Celebration". Computer History Museum. Retrieved September 13, 2008.

[17] Swenson, Reid C. (2007). "What is a Commodore Computer? A Look at the Incredible History and Legacy of the Commodore Home Computers". OldSoftware.Com. Retrieved November 19, 2007.

[18] http://sceneworld.org/blog/2015/02/12/video-interview-with-yash-terakura/

[19] Christopher Williams (4 August 2012). "Commodore 64 at 30: Computing for the Masses". *The Daily Telegraph*. Retrieved 16 August 2015.

[20] "PC – Model 5150". old-computers.com. Retrieved September 13, 2008.

[21] "Apple IIe". old-computers.com. Retrieved September 13, 2008.

[22] "Apple II+". old-computers.com. Retrieved September 13, 2008.

[23] "Atari 800". old-computers.com. Retrieved September 13, 2008.

[24] "Apple II History Chap 6". *Apple II History*. Retrieved November 17, 2014.

[25] "Commodore Commercials". commodorebillboard.de. Archived from the original on August 20, 2008. Retrieved September 13, 2008.

[26] Gupta, Anu M. (June 1983). "Commodore 64 Programmer's Reference Guide". *Compute!* (review). p. 134. Retrieved October 30, 2013.

[27] Tomczyk, Michael S. (1984). *The Home Computer Wars: An Insider's Account of Commodore and Jack Tramiel*. Compute! Publications. p. 110. ISBN 0942386787.

[28] "The CGW Computer Game Conference". *Computer Gaming World* (panel discussion). October 1984. p. 30.

[29] "Survey of Game Manufacturers". *Computer Gaming World*. April 1986. p. 32.

[30] Boosman, Frank (November 1986). "Designer Profiles / Alan Miller". *Computer Gaming World* (interview). p. 6.

[31] Protecto Enterprise (June 1983). "Commodore computer advertisement". *Popular Mechanics* (Hearst Magazines) **159** (6): p. 140. ISSN 0032-4558. We pack with your computer a voucher good for $100 rebate from the factory when you send in your old Atari, Mattel, Coleco electronic game or computer …

[32] Nocera, Joseph (April 1984). "Death of a Computer". *Texas Monthly* (Austin, Texas: Emmis Communications) **12** (4): pp. 136–139, 216–226. ISSN 0148-7736. Once before, Commodore had put out a product in a market where it chief competitor was TI: a line of digital watches. TI started a price war and drove Commodore out of the market. Tramiel was not about to let that happen again.

[33] Mitchell, Peter W. (1983-09-06). "A summer-CES report". *Boston Phoenix*. p. 4. Retrieved 10 January 2015.

[34] Remier, Jeremy. "A history of the Amiga, part 4: Enter Commodore". *arstechnica.com*. Retrieved August 4, 2008.

[35] Jacobs, Bob (January 1985). "An Agent Looks at the Software Industry". *Computer Gaming World*. p. 18.

[36] "Yle: Ohjelmat". November 20, 2011. Retrieved July 15, 2012.

[37] Wierzbicki, Barbara (1983-12-05). "Longevity of Commodore 64, VIC 20 questioned". *InfoWorld*. p. 24. Retrieved 13 January 2015.

[38] Halfhill, Tom R. (April 1986). "A Turning Point For Atari?". *Compute!*. p. 30. Retrieved November 8, 2013.

[39] Wagner, Roy (August 1986). "The Commodore Key". *Computer Gaming World*. p. 28. Retrieved November 1, 2013.

[40] "Compute! Gazette Issue 38".

[41] Lock, Robert; Halfhill, Tom R. (July 1986). "Editor's Notes". *Compute!*. p. 6. Retrieved November 8, 2013.

[42] Leemon, Sheldon (February 1987). "Microfocus". *Compute!*. p. 24. Retrieved November 9, 2013.

[43] Brooks, M. Evan (November 1987). "Titans of the Computer Gaming World / MicroProse". *Computer Gaming World*. p. 16.

[44] "Computer Chronicles: Interview with Commodore president with Max Toy". July 24, 2007. Retrieved July 24, 2007.

[45] Ferrell, Keith; Keizer, Gregg (September 1988). "Epyx Grows with David Morse". *Compute!*. p. 10. Retrieved November 10, 2013.

[46] Ferrell, Keith (July 1989). "Just Kids' Play or Computer in Disguise?". *Compute!*. p. 28. Retrieved November 11, 2013.

[47] Amiga Format News Special. "Commodore at CeBIT '94". Amiga Format, Issue 59 page 21, May 1994.

[48] Mike Holmes (15 April 2012). "Jack Tramiel and the Commodore 64". *Gamereactor*. Retrieved 6 August 2015.

[49] "The Educator 64 & Commodore PET 64 (aka C=4064)". zimmers.net. Retrieved September 13, 2008.

[50] "The 4064s: PET 64, Educator 64". School officials were dismayed at how easily the breadbox units could be stolen (in fact, quite a few disappeared from schools, and they fit very neatly in students' knapsacks), so Commodore presented the old PET cases as an inexpensive stopgap solution.

[51] "Secret Weapons of Commodore: The 4064s: PET 64, Educator 64". Retrieved November 17, 2014.

[52] Austin Modine (20 Jan 2008). "Remembering the Commodore SX-64". *The Register*. Retrieved 19 August 2015.

[53] "In Memory Of The Commodore C128 - Popular Science and Technology Blog by Jos Kirps". Retrieved November 17, 2014.

[54] Mace, Scott (January 28, 1985). "Commodore Shows New 128". *InfoWorld* (Menlo Park, CA: Popular Computing) **7** (4): pp. 19–20. ISSN 0199-6649.

[55] "Commodore C64GS". The Commodore 64 Games System, generally referred to as the C64GS is basically a Commodore 64 computer, with the keyboard and most other connectivity removed. You have the base unit, a cartridge port, two joystick ports, RF and Video outs... and that's your lot.

[56] John Markoff (20 December 2004). "A Toy with a Story". *The New York Times*. Retrieved 20 August 2015.

[57] Dunkels, Adam. "The Final Ethernet – C64 Ethernet Cartridge". Retrieved September 13, 2008.

[58] "SD2IEC on c64 wiki".

[59] "C64, C128 - Teil 2 - Retroport". retroport.de. June 14, 2013. Retrieved June 16, 2013.

[60] "Commodore 64: Web.It". amigahistory.co.uk. June 10, 2007. Retrieved June 16, 2013.

[61] "Iconic Commodore 64 All Set For Comeback".

[62] "Recreating the Legendary Commodore 64".

[63] "Commodore USA begins shipping replica C64s next week, fulfilling your beige breadbox dreams".

[64] "Commodore USA site showing assembly and boxed units ready for shipping".

[65] "C=4 Expo 2008". Lyonlabs.org. Retrieved April 22, 2013.

[66] Heimarck, Todd (June 1987). "When 2 + 3.5 + 4 = 7 / The Evolution of Commodore BASIC". *Compute!'s Gazette*. pp. 20–26. Retrieved June 30, 2014.

[67] "C64 Basic Introduction", *Commodore Magazine*, August 1982, p. 65.

[68] "Power C for the Commodore 64".

[69] "Wonderfully Ancient Aztec C Compilers".

[70] "Commodore BBS Outpost". Retrieved November 17, 2014.

[71] Alex Handy (23 September 2014). "The Strangest Software Project I've Ever Run". *SD Times*. Retrieved 19 August 2015.

[72] Ojala, Pasi. "Opening the Borders". Retrieved September 13, 2008.

[73] "New revolutionary C64 music routine unveiled". C64Music!. 2008. Retrieved May 20, 2014.

[74] Kirk, Mandy. "Commodore 64C System Guide at Auction on ebay". *www.ebay.com*. Commodore International. Retrieved October 17, 2014.

[75] Mace, Scott (November 13, 1983). "Commodore 64: Many unhappy returns". *InfoWorld* (Popular Computing Inc.) **5** (46): p. 23. ISSN 0199-6649.

[76] Anderson, John J. (March 1984). "Commodore". *Creative Computing*. p. 56. Retrieved 6 February 2015.

[77] Rautiainen, Sami. "Programmers_Reference". Retrieved March 23, 2011.

[78] Rautiainen, Sami. "Programmers_Reference". Retrieved March 23, 2011.

[79] MOS 6526 CIA datasheet (PDF format)

[80] Rautiainen, Sami. "Service_Manual: RAM Control Logic.". Retrieved March 13, 2011.

[81] "empty". 090505 computermuseum.li

[82] "The Hardware Book". Retrieved November 17, 2014.

[83] Carlsen, Ray. "C64 video port". Retrieved September 13, 2008.

[84] "250469 rev.A right". 100610 zimmers.net

[85] "250469 rev.A left". 100610 zimmers.net

[86] "Commodore C64 Power Supply Connector Pinout – AllPinouts". 090505 allpinouts.org

[87] "Commodore-64 BN/E 250469 schematic". 090519 zimmers.net

[88] "Commodore-64 BN/E 250469 schematic". 090519 zimmers.net

[89] "Commodore 64 memory map". sta.c64.org. February 4, 2013. Retrieved June 16, 2013.

[90] Wszola, Stan (July 1983). "Commodore 64". *BYTE*. p. 232. Retrieved October 20, 2013.

- Angerhausen, M.; Becker, Dr. A.; Englisch, L.; Gerits, K. (1983, 84). *The Anatomy of the Commodore 64*. Abacus Software (US ed.) / First Publishing Ltd. (UK ed.). ISBN 0-948015-00-4 (UK ed.). German original edition published by Data Becker GmbH & Co. KG, Düsseldorf.

- Bagnall, Brian (2005). *On the Edge: the Spectacular Rise and Fall of Commodore*. Variant Press. ISBN 0-9738649-0-7. See especially pp. 224–260.

- Commodore Business Machines, Inc., Computer Systems Division (1982). *Commodore 64 Programmer's Reference Guide*. Self-published by CBM. ISBN 0-672-22056-3.

- Tomczyk, Michael (1984). *The Home Computer Wars: An Insider's Account of Commodore and Jack Tramiel*. COMPUTE! Publications, Inc. ISBN 0-942386-75-2.

- Jeffries, Ron. "A best buy for '83: Commodore 64". *Creative Computing*, January 1983.

- Amiga Format News Special. "Commodore at CeBIT '94". *Amiga Format*, Issue 59, May 1994.

- Computer Chronicles; "Commodore 64 – Interview with Commodore president Max Toy", 1988.

- The C-64 Scene Database; "– Kjell Nordbø artist page (bio/release history) at CSDb".

- Michael Steil (December 29, 2008). *The Ultimate Commodore 64 Talk*. 25th Chaos Communication Congress (25c3). Berlin. Retrieved December 28, 2013. Lay summary.

1.1.9 External links

- Commodore 64: 8-Bit Legend 12 minute video compilation of Commodore 64 nostalgia and history

- C64 Preservation Project Discusses preservation of classic software for the Commodore 64

- Commodore 64 at DMOZ

- The History of the Commodore 64

- C64.com The home of Commodore 64 software and interviews with many game producers

- Commodore 64 history, manuals, and photos

- Extensive collection of information on C64 programming

- A History of Gaming Platforms: The Commodore 64 from October 2007

- Images of the C64 prototype from 2003

- A Commodore 64 Web Server Using Contiki v2.3

- Commodore 8-bit web links Project 70 sites and counting

- Commodore 64 Search Engine + Cloud

- "Commodore 64 still loved after all these years", *CNN*

- "The Commodore 64 at 25: thank you for the music", *The Guardian*

- Review of 5 different Commodore 64 models/motherboards, MOS6502.com

- Commodore Computer Club Discusses hardware, software, repairs and the scene for the Commodore 64

- "How to use the Commodore 64 for chiptunes", *2D-X*

- Variations on the Commodore 64 at the Wayback Machine (archived May 4, 2010)

- Commodore 64 turns 30: What do today's kids make of it?, *BBC*

- Design case history: the Commodore 64, IEEE Spectrum, March 1985

- Kernal64: A Commodore 64 Emulator in Java/Scala

- Commodore 64 Game scores

Chapter 2

Further Reading (Alphabetical Order)

2.1 1541 Ultimate

1541 Ultimate Plus

1541 Ultimate (often abbreviated *1541U*) is a peripheral, primarily an emulated floppy disk and cartridge emulator based on the FPGA Xilinx XC3S250E, for the home computer Commodore 64 (C64). It became available in 2008.

The unit is developed by Gideon Zweijtzer and is a C64-compatible cartridge that among other things contains older cartridges like Action Replay, The Final Cartridge III, Super Snapshot, Retro Replay or TurboAss with Codenet-support, and a mostly fully compatible FPGA-cloned Commodore 1541 floppy disc unit that can use C64-compatible files like .D64 disc images or .PRG files via a SD-card reader. The version plus has 32 MB RAM. It's possible to connect Ethernet to the unit.

In 2010 the 1541 Ultimate II was developed, that is smaller than the original and that has certain hardware differences like USB interface beyond MicroSD. In addition, all firmware and VHDL code for the Ultimate II is licensed under an open source license, specifically the GPLv3, allowing hobbyists and others to freely modify all aspects of its functionality, including the FPGA-emulated hardware.

A similar peripheral for the same need is the MMC64 developed in 2005.

2.1.1 See also

- Commodore 64
- Commodore 1541
- Commodore 64 peripherals

2.1.2 External links

- 1541ultimate.net - 1541 Ultimate homepage

2.2 Action Replay

For the Indian film, see Action Replayy. For the Howard Jones EP, see Action Replay (EP).

Action Replay is the brand name of a series of video-

Action Replay cartridge for the Amiga 500

game cheating devices (such as cheat cartridges) created by Datel. Action Replays are available for major modern gaming systems including the Nintendo DS, Nintendo DSi, and

Action Replay cartridge for Commodore 64

Action Replay MAX DUO for Nintendo DS

the PlayStation Portable, as well as older systems including the PlayStation 2, Nintendo GameCube, Game Boy Advance, and the Xbox. PowerSaves by Action Replay is a related series of video-game cheat devices that store game saves created by Datel, which allow users to cheat without modifying the game code being executed—unlike the main

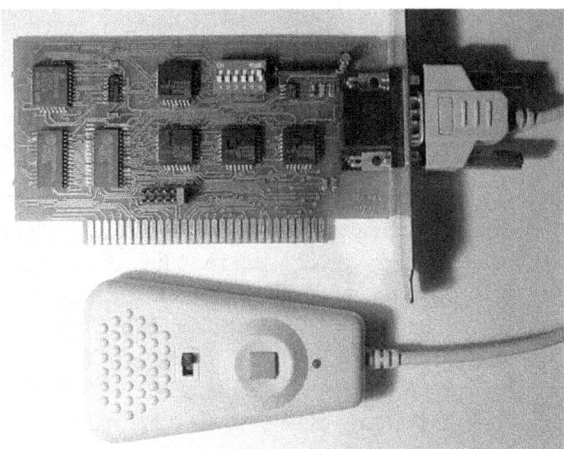

Action Replay ISA card for PC 1994

Action Replay series, which cheats by modifying game code itself. PowerSaves are available for game systems such as the Wii on an SD card and the 3DS.

2.2.1   Typical features

Typically options include cheats, level warping, and display of internal game data not normally viewable by the player.

- Infinite lives, invulnerability, permanent power-ups, no collision detection, walk through walls, one-hit kills, super-high jumps, infinite money, etc.

- Obtain any item in the game, even those not normally obtainable (e.g. debug or removed items).

- Access or warp to any level, even those not normally accessible (e.g. test or unused levels).

- Activate debug menus, normally used by programmers when testing and debugging a game.

- Download, upload, import and export save games to the Internet or a storage device.

- Save game state to disk, so it can be restarted from that point even if the game does not support saving.

- Region-free operation.

- Loading of third-party homebrew applications/games, not backup copies of retail games.

2.2.2   Criticisms

Datel, the maker of Action Replay, has received several criticisms from the gaming world over its products. One of the most frequent complaints is the so-called "planned ob-solescence", where codes for a just-released game require the most recent version of the cheat software.

Datel encrypted the codes on the Action Replay for PS2, GC and GBA. The encryption was meant to stop hackers from translating its codes for use in other cheating devices, but it also prevented users from making their own codes, as well as the creation of codes using a template. (There is, however, a program called 'GCNCrypt' that decrypts and encrypts Action Replay codes for the Nintendo GameCube, making editing and hacking of codes possible.) Cheat codes normally involve a memory address, a value, and some-times a trigger that says when the code is activated (always on, on at the start, on after a certain button press). Some-times, cheat codes possess a pointer address which points to multiple other memory addresses. Therefore, for some games, it is possible to create a code template and derive hundreds of codes by modifying the values. For example, in a role-playing game, one can use a code template and a table of values to create a code that will give any charac-ter any piece of equipment in the game. With encrypted codes, it was not possible to use such a template, and any codes had to be created and distributed by Datel. However, because of the sheer number of codes that can be created in this fashion, it was not plausible for Datel to release a list of codes with this versatility. Action Replay for the DS now allows cheat codes (the previous Action Replay only managed game saves), using unencrypted codes, and has a trainer toolkit available that allows users to create their own codes.

The PS2 Action Replay occasionally corrupts the memory cards, leaving corrupt files on the card that cannot be deleted by the PS2. The Action Replay can fix the memory card by formatting it, but the corrupted data cannot be restored.

Cheating in online games is also usually frowned upon, with game companies making efforts to prevent and detect it. However, with an Action Replay it is possible to cheat with-out being detected, or in a game for which there is normally no way to cheat. Examples include *Phantasy Star Online* for the Dreamcast, in which it was possible to manufacture items offline using an Action Replay and then carry them online undetected; there was no way to determine if an item had been manufactured or legitimately won. Action Replay can also disable anti-cheating code and prevent detection; however, since most modern versions only allow codes to be created by Datel and they have so far not taken this route, there are no anti-detection codes—for current-generation systems.

Other criticisms include the loss of data and/or progress when using Action Replay. Sometimes, data loss can make a game unavailable to play. Entering an inappropriate or wrong code may not cause a noticeable loss of play until

Action Replay is removed from the user's system, at which point the data error takes effect.

2.2.3 Versions for computers

- Commodore 64

  - Action Replay
  - Action Replay MK II
  - Action Replay MK III
  - Action Replay MK IV (1988)
  - Action Replay MK V (1989)
  - Action Replay MK VI

- Commodore Amiga

  - Action Replay (A500 cart / A2000 CPU card)
  - Action Replay (A1200 card)
  - Action Replay MK II (A500 cart / A2000 CPU card)
  - Action Replay MK III (A500 cart / A2000 CPU card) (1991)

- PC

  - Action Replay PC (ISA card) for DOS (1994)
  - Action Replay PC for Windows 95/98 (1998)

The ISA-based Action Replay needed memory-resident drivers for real and protected mode. The card had a grab-ber, a trainer, and a slowdown feature. It could also inter-rupt the current game or save it to disk (freezer).

Models running firmware 4.0 and beyond use EEPROM in-stead of ROM and thus are upgradeable.[1]

In December 1998, Datel released a version for Windows 95/98.[2]

2.2.4 Versions for video game consoles

Third generation

- Nintendo Entertainment System

  - Pro Action Replay

- Sega Master System

  - Pro Action Replay

Fourth generation

- Sega Mega Drive/Sega Genesis

  - Action Replay
  - Pro Action Replay
  - Pro Action Replay MK2
  - Pro CDX (Action Replay) for the Mega-CD

- Super Nintendo Entertainment System

  - Pro Action Replay
  - Pro Action Replay MK2
  - Pro Action Replay MK3

Fifth generation

- Sega Saturn

  - Pro Action Replay
  - Pro Action Replay 4M (with 4 MB RAM)
  - Pro Action Replay 4M Plus (Same as the 4M, but with manual choice of the needed RAM)

- PlayStation

  - Action Replay (1995)
  - Pro Action Replay (1996)
  - Action Replay CDX (1997)
  - Action Replay 2 V2 (2001) [As Bonus Disc With PS2 Action Replay 2 V2]
  - Equalizer
  - Equalizer CDX
  - Equalizer Xtreme

- Nintendo 64

  - Action Replay
  - Action Replay Professional (1999)
  - Equalizer

Sixth generation

- Dreamcast

  - Action Replay CDX (2000)
  - Equalizer Xtreme

- PlayStation 2

  - Action Replay 2 (2000)
  - Action Replay 2 V2 (2001)

- Action Replay MAX (2003)
- Action Replay MAX EVO (2004)
- Action Replay MAX EVO (2009)

- Xbox

  - Action Replay (2002)
  - Action Replay MAX
  - Action Replay MAX 360 Powersaves (2009)

- Nintendo GameCube

  - Action Replay (2003) [Note: The latest Wii firmware blocks this on Wii consoles running in GameCube mode.]
  - Action Replay MAX (200X)
  - Action Replay (2006, works on Wii)
  - Action Replay Powersaves (2007)

Seventh generation

- Wii

  - Wii Action Replay Powersaves (2010)
  - Action Replay Wii (2012)

2.2.5   Versions for hand-held consoles

- Sega Game Gear

  - Pro Action Replay

- Game Boy, Game Boy Pocket, Game Boy Color

  - Pro Action Replay
  - Action Replay Professional (1997)
  - Action Replay Pro (1999)
  - Action Replay Online (2000)
  - Action Replay Xtreme (2001)

- Game Boy Advance, Game Boy Advance SP, Game Boy Micro

  - Action Replay GBX (November 2001)
  - Action Replay (2003)
  - Action Replay MAX (2004)
  - Action Replay MAX DUO (March 2005)

- Nintendo DS, Nintendo DS Lite

  - Action Replay MAX DUO (March 2005)

- Action Replay DS (July 2006) [Last firmware v1.71, games released later are not compatible]

  - NDS Trainer Toolkit (February 2007) [available only online] Toolkit Manual

  - Action Replay DS Media Edition (September 2008) [available only online]

  - Action Replay DS EZ (February 2009)

- Nintendo DSi, Nintendo DSi XL

  - Action Replay DSi (October 2009); Later system software updates to the DSi and 3DS include a 'white list' which prevents unlicensed games from booting,[3] this stops older Action Replay's from loading on updated DSi and 3DS handhelds, however older Action Replay's will continue to work on original DS and DS Lite handhelds.

  - Action Replay DS "3DS/DSi/DS/Lite Compatible" (September 2011)

- Nintendo 3DS, Nintendo 3DS XL

  - Action Replay Power Saves for 3DS (June 2013). The Action Replay Power Saves for 3DS can alter saves of 3DS games and has some codes for 3DS games.

- PlayStation Portable

  - Action Replay MAX including 64 MB Memory Stick (August 2005) [Powersaves only]

  - Action Replay for PSP including 64 MB or 1 GB Memory Stick [Powersaves only]

  - Action Replay PSP including 1 GB Memory Stick (October 2008)

  - Action Replay PSP Online (December 2009)

2.2.6 See also

- GameShark

- Game Genie

- Multiface

- Code Breaker

- Xploder

2.2.7 References

[1] "README included in ZIP archive for PC". 2009-12-11. Retrieved 2009-12-11.

[2] "statement retrieved from Archive.org". 2009-12-11. Retrieved 2009-12-11.

[3] "Datel Lawsuit coming in 3.. 2.. 1..". 2010-02-01. Retrieved 2014-03-08.

2.2.8 Links

- Official Action Replay website

- Official Action Replay Powersaves for 3DS Website

2.3 C-One

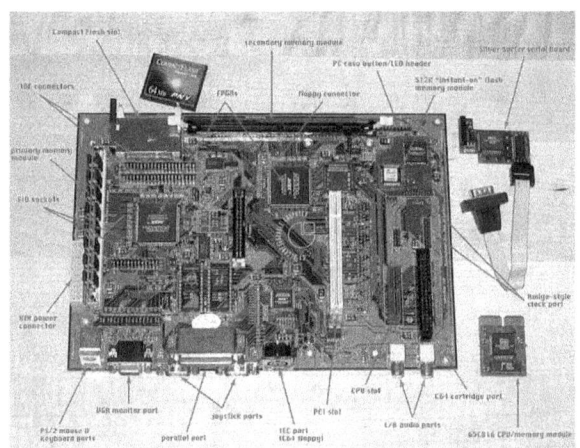

The C-One

The **C-One** is a single-board computer (SBC) created in 2002 as an enhanced version of the Commodore 64, a home computer popular in the 1980s. Designed by Jeri Ellsworth, a self-taught designer, and Jens Schönfeld from Individual Computers, who manufactured the boards themselves, the C-One has been re-engineered to allow cloning of other 8-bit computers.

2.3.1 Design

The machine uses a combination of configurable Altera field-programmable gate array (FPGA) chips and modular CPU expansion cards to create compatibility modes that duplicate the function of many older home computers. The default CPU is the 65C816 CPU which is used in Commodore 64 compatibility mode as well as the C-One's native operating mode. The C-One is not merely a software emulator, it loads various "core" files from disk to configure the FPGA hardware to recreate the operation of the core logic chipsets found in vintage computers. This provides for a very accurate and customizable hardware emulation platform. The C-One is not limited to recreating historical computers: its programmable core logic can be used to create entirely new custom computer designs.

In 2004, the platform was expanded to include an Amstrad CPC core made by Tobias Gubener.

In 2006, Peter Wendrich ported his FPGA-64 project (originally intended for a Xilinx FPGA) and enhanced it for the C-One.[1] This core supported both PAL and NTSC machine emulation, and aimed to be cycle-exact and emulate many of the bugs and quirks of the original hardware.

In 2008, after development of an "Extender" card which added a third FPGA, Tobias Gubener added Amiga 500 compatibility by porting Dennis van Weeren's Minimig code to the board.[2] This core replaced the physical 68000 CPU and the PIC chip from the original with his own TG68 CPU core on the FPGA. Recent developments to this core include features not possible with the original Minimig board.[3]

In 2009, Peter Wendrich released a "preview" of a next-generation C64 core called "Chameleon 64", with a greatly expanded specification compared to his earlier core.[4] A new version of the CPC core was also released in mid-2009, featuring an embedded SymbOS core for control of device emulation, and a clock unlocked mode for CPU speeds of up to 80 MHz.

So far, C-One circuit boards have been produced by German company Individual Computers, and they currently sell for €333 with the FPGA extender card.

2.3.2 See also

- C64 Direct-to-TV

- Sprinter (computer)

- 1chipMSX

- Home computer remakes

2.3.3 References

[1] http://www.syntiac.com/fpga64.html

[2] http://www.jschoenfeld.com/news/news133_e.htm

[3] http://www.c64upgra.de/c-one/s_news.htm

[4] http://www.syntiac.com/chameleon.html

2.3.4 External links

- Official website

- Wiki

- Retroputing's forum on the C-One

- Yahoo's C-One Group

- Yahoo group for technical discussion about development of cores for the C-One

2.4 C64 Direct-to-TV

The **C64 Direct-to-TV**, called **C64DTV** for short, is a single-chip implementation of the Commodore 64 computer, contained in a joystick (modeled after the mid-1980s Competition Pro joystick), with 30 built-in games. The design is similar to the Atari Classics 10-in-1 TV Game. The circuitry of the C64DTV was designed by Jeri Ellsworth, a self-taught computer chip designer who had formerly designed the C-One.

The C64 Direct-to-TV computer-in-a-joystick unit.

Tulip Computers (which had acquired the Commodore brand name in 1997) licensed the rights to Ironstone Partners, which cooperated with DC Studios, Mammoth Toys, and "The Toy:Lobster Company" in the development and marketing of the unit.[1] QVC purchased the entire first production run of 250,000 units and sold 70,000 of them on the first day that they were offered.

2.4.1 Versions

There exist multiple versions of the C64DTV. DTV1 (NTSC television type) comes with 2 MB ROM. It first appeared in late 2004 for the American/Canadian market. DTV2 (called *C64D2TV* sometimes) is a revised version for the European and world markets (PAL television type) and appeared in late 2005. The ROM has been replaced by flash memory in these devices. However, the DTV2/PAL version suffers from a manufacturing fault, which results in poor colour rendering (the resistors in the R-2R ladder DACs for both the chroma and the luma have been transposed). In the DTV3, a problem with the blitter was fixed.

2.4.2 Hardware Specifications

Commodore DTV PCB.

- Core circuity
  - ASIC running at 32 MHz internally, implementing 6510 CPU, VIC-II, SID, CIA, and PLA
- Casing/Connectors
  - integrated in a joystick (as if connected to port 2 of a real C64)
  - five additional buttons (acting like keys)

- running from batteries only (four AA batteries)
- Composite video, monaural audio (RCA connectors)
- looks similar to a *Competition Pro* joystick

- Graphics
  - NTSC (DTV2 and later: NTSC/PAL on chip, only PAL wired in end-market devices)
  - reprogrammable palette with 4 bits of luma and 4 bits of chroma
  - DTV2 and later: "chunky" 256 color mode, additional blitter for fast image transformation

- Sound
  - no support for SID filters
  - DTV2 and later: 8 bit digital sound, additional options for envelope generators

- Memory
  - DTV1: 128 KB RAM, 2 MB ROM
  - DTV2 and later: 2 MB RAM, 2 MB flash memory
  - DMA engine for RAM/RAM and ROM/RAM transfers
  - DTV2 and later: additional RAM access using bank switching and blitter

- CPU
  - implementing a 6510 at 1 MHz
  - DTV2 and later: Enhanced CPU (fast/burst mode, additional registers and opcodes, support for illegal ops of the 6510)

2.4.3 Built-in games

The official games for the unit are mostly a mix of Epyx and Hewson C64 games. Games unique to the NTSC or PAL versions are noted below.

2.4.4 Hardware-modding

Since the internal circuit board has exposed solder points for floppy-drive and keyboard ports, hardware modifications of the C64DTV are relatively simple.

Known hardware mods

- keyboard connector

- external joystick (Port 1 and 2)

- floppy connector

- power unit connector

- fixing the palette problems of the PAL version (to some degree this is possible in software by adjusting palette entries)

- S-Video connector

- user port

- Original C64 casing and PS2 keyboard [2]

Additional hardware

- Data transfer cable (Parallel port (or USB/serial port via DTV2ser) to Joystick or user port)

- SD card interface *1541-III* or *MMC2IEC*

2.4.5   Limitations

The internal flash memory is accessible as device 1. However, software is not included to support write operations so high score saving is not possible. Also, flash devices used in the DTV are specified for a very limited number of write accesses only.

When using the standard keyboard mod, the F7 key does not work. There is a workaround, the "Keyboard Twister."[3]

2.4.6   Software-modding

The DTV contains software-flashable memory. A number of tools have been released to compile programs into DTV-compatible flash images and load it onto the DTV. People made their own game compilations, adding popular (sometimes DTV-fixed) games that were not in the original DTV, added boot menus to make homebrew software development easier or enable new features, for example transfer programs like DTVtrans for transferring data from PC to DTV RAM and vice versa via the PC parallel port (or USB) and the DTV joystick port.

2.4.7   References

[1] The Commodore 64 bounces back to life as a Direct-To-TV plug and play Joystick! // GamesIndustry.biz

[2] "C64DTV in original C64 case". Joco.homeserver.hu. Retrieved 2011-07-19.

[3] "Keyboard Twister by Shadowolf". Picobay.com. 2009-10-02. Retrieved 2011-07-19.

2.4.8   External links

- *DTV Hacking Wiki*, archived from the original on 2013-04-14, retrieved 2013-08-06 - DTV versions overview, HOWTOs, DTV Programming guide

- The Official C64 DTV site - user manual plus some other information

- David Murray's Commodore DTV Hacking

- C64DTV stuff by tlr Flash Tool, ML-Monitor, PC<->DTV transfer system

- Mr. Latch-up's C64 DTV & Hummer Advice Column

- A page about the history of the device

- Details on fixing colour problem on PAL DTVs - Note that surface-mount soldering skills are required.

- DTVtrans, connecting a DTV to a PC via parallel port

- DTV2ser, connecting a DTV to a PC/Mac via USB or serial port

- Four ways to turn a C64 DTV into a C64 clone

- Grokk´s DTV Stuff DTVBIOS and DTVBASIC - make your DTV code-ready.

2.5   CARDCO

CARDCO was a computer peripheral company during the 1980s in Wichita, Kansas, United States. CARDCO was well known in the Commodore 64 and VIC-20 community because of advertisements in numerous issues of Compute! magazine and availability of their products at large retailers.[1]

There were severe shortcomings of early Commodore printers, so CARDCO created the Card Print A (C/?A) printer interface that emulated Commodore printers by converting the Commodore-style IEEE-488 serial interface to a Centronics printer port to allow numerous 3rd-party printers to be connected to a Commodore 64 or VIC-20, such as Epson, Okidata, C. Itoh.[2] A second model, a version that supported printer graphics was released called the Card Print +G (C/?+G), supported printing Commodore graphic characters using ESC/P escape codes. CARDCO released additional enhancements, including a model with RS-232 output, and shipped a total over two million printer interfaces.

2.5.1   See also

- Commodore 64 peripherals

2.5.2 References

[1] *Compute!* on Internet Archive

[2] "CARDCO Card Print A (C/?A) - Printer Interface For The Commodore 64 and VIC-20". *COMPUTE Magazine* (34): 251. March 1983.

2.5.3 External links

- CARDCO Card Print A (C/?A) Printer Interface: User Manual, Addendum

- CARDCO Card Print +G (C/?+G) Printer Interface: User Manual, Supplement

2.6 cc65

cc65 is a complete cross development package for 65(C)02 systems, including a powerful macro assembler, a C compiler, linker, librarian and several other tools.

It is based on a C compiler that was originally adapted for the Atari 8-bit computers by John R. Dunning. The original C compiler is a Small C descendant but has several extensions, and some of the limits of the original Small C compiler are gone.

The toolkit has largely been expanded by Ullrich von Bassewitz and other contributors. The actual cc65 compiler, a complete set of binary tools (assembler, linker, etc.) and runtime library are under a license identical to zlib's.[1] The *ca65* cross-assembler is one of the most powerful 6502 cross-assemblers available under an open-source license.

The compiler itself is almost completely ANSI C compatible, though not completely. The C library is quite extensive, and allows extensive usage of the target platform's hardware. stdio is supported on many platforms, as is Borland-style conio.h screen handling. GEOS is also supported on the Commodore 64 and even the Apple II. The library supports many of the Commodore platforms (C64, C128, C16/116/Plus/4, P500 and 600/700 family), Apple II family, Atari 8-bit family, Oric Atmos, Nintendo Entertainment System and Watara Supervision game console.

The officially supported host systems include Linux, Microsoft Windows, DOS and OS/2, but the source code itself is quite portable and has been reported to work almost unmodified on many platforms beside these.

2.6.1 Supported API

static

- conio (text-based console I/O non-scrolling)

- dio (block-oriented disk I/O bypassing the file system)

dynamic

- em (expanded memory, used for all kinds of memory beyond the 6502's 64K barrier, similar EMS)

- joystick (relative input devices)

- mouse (absolute input devices)

- serial (communication)

- tgi (2D graphics primitives inspired by BGI)

Note: For static libraries, "Yes" means the feature is available. For dynamic libraries, the columns list the number of available drivers.

[1] https://github.com/cc65/cc65/commit/aeb849257277a6b98542de8579697b81c6dd70e6

[2] By Fatih Aygün. CIRCLE doesn't work at all, some graphics modes may crash on some machines.

2.6.2 External links

- Official website (no longer maintained)
- Modern github fork of cc65
- Contiki desktop, written with cc65
- TGI drivers for atari8
- Atari TGI 2009-11-02 release announcement on cc65 mailing list

2.7 CMD RAMLink

The **RAMLink** was one of several RAM expansion products made by Creative Micro Designs (CMD) for Commodore's C64/128 home computers. The RAMLink was intended as a third-party alternative, successor and optionally companion to Commodore's own 17xx-series REU RAM expansion cartridges.

Unlike the REU, the RAMLink is externally powered and designed from the ground-up to act as a RAM disk.

RAMLink device

2.7.1   Features

- Allows up to 16 MB of expansion RAM. The expansion memory can be provided by a combination of 30-pin SIMM RAM on an internal card, a Commodore 17xx-series REU (or a clone) plugged into the RAM Port, or a GeoRAM.

- Provides its own copy of JiffyDOS, allowing accelerated operation with any other JiffyDOS-equipped disk device, as well as shorthand commands (DOS Wedge) to conveniently access any other connected storage devices.

- Full set of partitioning tools and DOS commands.

- Commodore 1541, 1571 and 1581 disk-layout emulation modes

- One partition type provides Direct-access REU-like capability

- Secondary power socket and on-board charging circuit to accept a 6-volt "sealed" lead acid backup battery.

- Battery-backed real time clock for time and date stamping of files, if the internal RAM card is present.

- Includes drivers to allow GEOS to use its memory as either a replacement for swap space, or as a regular 'disk' drive.

- Custom parallel connection for the CMD HD Series line of hard drives.

- Pass-through expansion port for standard cartridges (e.g. Action Replay, Super Snapshot)

- On-device buttons to swap device numbers with other drives, switches to disable 17xx-series REUs or change their handling.

2.7.2   References

- Creative Micro Designs (1990). *CMD RAMLink User's Manual, third edition.* (supplied with the hardware)

2.7.3   External links

- Technical info, Games & Utilities

2.8   Commodore 64 peripherals

Commodore 64 Home Computer

This article is about the various external peripherals of the Commodore 64 home computer.

2.8.1   Storage

Tape drives

Main article: Datasette

In the United States, the 1541 floppy disk drive was widespread. By contrast, in Europe, the C64 was often used with cassette tape drives (Datasette), which were much cheaper, but also much slower than floppy drives. The Datasette plugged into a proprietary edge connector on the Commodore 64's motherboard. Standard blank audio cassettes could be used in this drive. Data tapes could be write-protected in the same way as audio cassettes, by punching out a tab on the cassette's top edge.

The Datasette's speed was very slow (about 300 baud). Loading a large program at normal speed could take up to 30 minutes in extreme cases. Many European software developers wrote their own fast tape-loaders which replaced the internal KERNAL code in the C64 and offered loading times more comparable to disk drive speeds. **Novaload** was perhaps the most popular tape-loader used by British and American software developers. Early versions of **Novaload** had the ability to play music while a program loaded into

Commodore Datasette 1530

5.8 kB/s and included "freezer" capabilities.[3]

Floppy disk drives

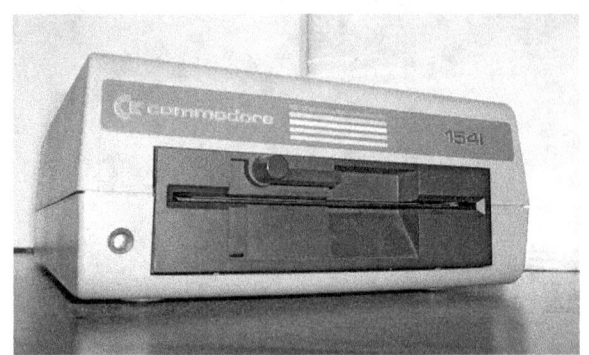

Commodore 1541 Floppy Drive

memory, and was easily recognizable by its black border and digital bleeping sounds on loading. Other fast-loaders included load screens, displaying computer artwork while the program loaded. More advanced fast-loaders included minigames for the user to play while the program loaded from cassette. One such minigame fastloader was Invade-a-Load.

Users also had to contend with interference from magnetic fields. Also, not too dissimilar to floppy drive users, the Datasette's read head could become dirty or slip out of alignment. A small screwdriver could be used to align the tape heads, and a few companies capitalized by selling various commercial kits for Datasette head-alignment tuning.

As the Datasette lacked any random read-write access, users had to either wait while the tape ran its length, while the computer printed messages like "SEARCHING FOR ALIEN BOXING... FOUND AFO... FOUND SPACE INVADERS... FOUND PAC-MAN... FOUND ALIEN BOXING... LOADING..." or else rely on a tape counter number to find the starting location of programs on cassette. Tape counter speeds varied over different datasette units making recorded counter numbers unreliable on different hardware.

An optional streaming tape drive, based upon the QIC-02 format, was available for the Xetec Lt. Kernal hard drive subsystem (see below). They were expensive and few were ever sold.

A similar concept to the ZX Microdrive (85 kB) was the extremely fast "**Phonemark 8500 Quick Data Drive**" which has 128 kB capacity using a micro-cassette storage unit and used the C2N Datasette. The concept eventually succumbed to floppy drives.[1][2]

Backup to VHS tapes were offered by **DC Electronics** with their cartridge **WHIZZARD** in 1988. Which could handle

Although usually not supplied with the machine, floppy disk drives of the 5¼ inch (1541, 1570 and 1571) and, later, 3½ inch (1581) variety were available from Commodore.

The 1541 was the standard floppy disk drive for the Commodore 64, with nearly all disk-based software programs released for the computer being distributed in the 1541 compatible floppy disk format. The 1541 was very slow in loading programs because of a poorly implemented serial bus, a legacy of the Commodore VIC-20.

The 1541 disk drive was notorious for not only its slow performance and large physical size compared to the C64 (the drive is almost as deep as the computer is wide), but also for the drive mechanisms installed during early production runs, which quickly gained a bad reputation for their mechanical unreliability.

Perhaps the most common failure involved the drive's read-write head mechanism losing its alignment. Due to lack of hardware support for detecting track zero position, Commodore DOS formatting routines and many complex software copy-protection schemes (which used data stored on nonstandard tracks on floppies) had to rely on moving the head specified number of steps in order to make sure that the desired head position for formatting or reading the data was reached. Since after physically reaching track zero, further movement attempts caused the head drive mechanism to slam (producing the infamous, loud, telltale knocking sound) into a mechanical stop, the repetitive strain often drove the head mechanism out of precise alignment, resulting in read errors and necessitating repairs. As a side note: some demos exploited the sound generated by the head moving stepper motor to force the disk drive to play crude tunes ("Bicycle Built For Two" was one) by varying the frequency of step requests sent to the motor.

Also, as with the C64, 1541 drives tended to overheat due

to a design that did not permit adequate cooling (potentially fixed by mounting a small fan to the case). Many of the 1541's design problems were eventually rectified in Commodore's 1541-II disk drive, which was compatible with the older units. The power supply unit was not housed inside the drive case; hence, the 1541-II size was significantly smaller and did not overheat.

Because of the drive's initial high cost (about as much as the computer itself) and target market of home computer users, BASIC's file commands defaulted to the tape drive (device 1). In order to load a file from a commercial disk, the following command must be entered:

LOAD "*",8,1

In this example, '*' designates the last program loaded, or the first program on the disk, '8' is the disk drive device number, and the '1' signifies that the file is to be loaded not to the standard memory address for BASIC programs, but to the address where its program header tells it to go—the address it was saved from. This last '1' usually signifies a machine language program.

Commodore 1541C Floppy Drive, 2nd model

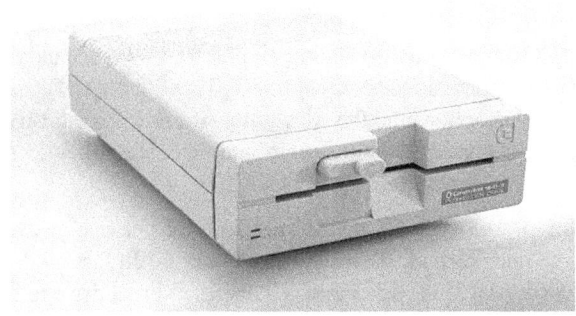

Commodore 1541-II Floppy Drive, 3rd model

Not long after the 1541's introduction, third-party developers demonstrated that performance could be improved with software that took over control of the serial bus signal lines and implemented a better transfer protocol between the computer and disk. In 1984 Epyx released its *FastLoad* cartridge for the C64, which replaced some of the 1541's slow routines with its own custom code, thus allowing users to load programs in a fraction of the time. Despite being incompatible with many programs' copy protection schemes, the cartridge became so popular among grateful C64 owners (likely the most-widespread third-party enhancement for the C64 of all time) that many Commodore dealers sold the Epyx cartridge as a standard item when selling a new C64 with the 1541.

As a free alternative to FastLoad cartridges, numerous pure software *turbo-loader* programs were also created that were loaded to RAM each time after the computer was reset. The best of these turbo-loaders were able to accelerate the time required for loading a program from the floppy drive by a factor of 20x, demonstrating the default bus implementation's inadequacy. As turbo-loader programs were relatively small, it was common to place one on almost each floppy disk so that it could be quickly loaded to RAM after restart.

The 1541 floppy drive contained a MOS 6502 processor acting as the drive controller, along with a built-in disk operating system (DOS) in ROM and a small amount of RAM, the latter primarily used for buffer space. Since this arrangement was, in effect, a specialized computer, it was possible to write custom controller routines and load them into the drive's RAM, thus making the drive work independently of the C64 machine. For example, certain back up software allowed users to make multiple disk copies directly between daisy-chained drives without a C64.

Several third party vendors sold an IEEE-488 general purpose interface bus adapter for the C64, which plugged into the machine's expansion port. Outside of BBS operators, few C64 owners took advantage of this arrangement and the accompanying IEEE devices that Commodore sold (such as the SFD-1001 1-megabyte 5¼ inch floppy disk drive, and the peripherals originally made for the IEEE equipped PET computers, such as the 4040 and 8050 drives and the 9060/9090 hard disk drives).

As an alternative to the feeble performing 1541 or the relatively expensive IEEE bus adapter and associated peripherals, a number of third-party serial-bus drives such as the MSD Super Disk and Indus GT appeared that often offered better reliability, higher performance, quieter operation, or simply a lower price than the 1541, although often at the expense of software compatibility due to the difficulty of reverse engineering the DOS built into the 1541's hardware (Commodore's IEEE-based drives faced the same issue due to the dependence of the DOS on features of the Commodore serial bus).

Like the IEEE-488 interface, the serial bus offered the ability to daisy chain hardware together. This led to Commodore producing (via a third party) the Commodore 4015, or VIC-switch. This device (now rarely seen) allowed up to 8 Commodore 64s to be connected to the device along with a string of peripherals, allowing each computer to share the connected hardware.

It was also possible, without requiring a VIC-switch, to connect two Commodore 64s to one 1541 floppy disk drive to simulate an elementary network, allowing the two computers to share data on a single disk (if the two computers made simultaneous requests, the 1541 admirably handled one whilst returning an error to the other, which surprised many people who expected the 1541's less-than-stellar drive controller to crash or hang). This functionality also worked with a mixed combination of PET, VIC-20, and other selected Commodore 8-bit computers.

In the mid-1980s, a 2.8-inch floppy disk drive, the Triton Disk Drive and Controller, was introduced by Radofin Electronics, Ltd. It was compatible with the Commodore 64 as well as other popular home computers of the time, thanks to an operating system stored on an EPROM on an external controller. It offered a capacity of 144/100 kilobytes non-formatted/formatted, and data transfer rates of up to 100 kilobytes per second. Up to 20 files could be kept on each side of the double-sided floppy disks.

Later in the 1990s, Creative Micro Designs produced several powerful floppy disk drives for the Commodore 64. These included the FD-Series serial bus compatible 3.5″ floppy drives (FD-2000, FD-4000), which were capable of emulating Commodore's 1581 3.5″ drive as well as implementing a native mode partitioning which allowed typical 3.5″ high-density floppy disks to hold 1.6 MB of data—more than MS-DOS's 1.4 MB format. The FD-4000 drive had the advantage of being able to read hard-to-find enhanced floppy disks and could be formatted to hold 3.2 MB of data. In addition, the FD series drives could partition floppy disks to emulate the 1541, 1571 and 1581 disk format (although unfortunately, not the emulated drive firmware), and a real time clock module could be mounted inside the drive to time-stamp files. Commercially, very little software was ever released on either 1581 disk format or CMD's native format. However, enthusiasts could use this drive to transfer data between typical PC MS-DOS and the Commodore with special software, such as SOGWAP's Big Blue Reader.

There was one other 3.5″ floppy drive available for the Commodore 64. The "TIB 001" was a 3.5″ floppy drive that connected to the Commodore 64 via the expansion port, meaning that these drives were very fast. The floppy disks themselves relied on an MS-DOS disk format, and being based on cartridge allowed the Commodore 64 to boot from them automatically at start-up. These devices

appeared from a company in the United Kingdom, but unfortunately did not become widespread due to non-existent third-party support. In an article in *Zzap!64* of November 1991, several software houses interviewed believed that the device came to the market too late to be worthy of supporting.

Hard drives

Seagate ST 506 5¼-inch HDD with cover removed.

Late in 1984, Fiscal Information Inc., of Florida, demonstrated the Lt. Kernal hard drive subsystem for the C64. The Lt. Kernal mated a 10 megabyte Seagate ST-412 hard drive to an OMTI SASI intelligent controller, creating a high speed bus interface to the C64's expansion port. Connection of the SASI bus to the C64 was accomplished with a custom designed host adapter. The Lt. Kernal shipped with a disk operation system (DOS) that, among other things, allowed execution of a program by simply typing its name and pressing the Return key. The DOS also included a keyed random access feature that made it possible for a skilled programmer to implement ISAM style databases.

By 1987, the manufacturing and distribution of the Lt. Kernal had been turned over to Xetec, Inc., who also introduced C128 compatibility (including support for CP/M). Standard drive size had been increased to 20 MB, with 40 MB available as an option, and the system bus was now the industry-standard small computer system interface, better known as SCSI (the direct descendant of SASI).

The Lt. Kernal was capable of a data transfer rate of over 38 kB per second (65 kB per second in C128 fast mode). An optional multiplexer allowed one Lt. Kernal drive to be shared by as many as sixteen C64s or C128s (in any combination), using a round-robin scheduling algorithm that took advantage of the SCSI bus protocol's ability to handle multiple initiators and targets. Thus the Lt. Kernal could be conveniently used in a multi-computer setup, something that was not possible with other C64-compatible hard drives.

Production of the Lt. Kernal ceased in 1991. Fortunately, most of the components used in the original design were industry standard parts, making it possible to make limited repairs to the units. In 2010, a re-creation of the Lt. Kernal was produced by MyTec Electronics. It was called the Rear Admiral HyperDrive and used an upgraded DOS called RA-DOS. The Rear Admiral parts could be used to upgrade the older Lt. Kernal, e.g. chips from the Rear Admiral host adapter could be used to upgrade the chips in the Lt. Kernal host adapter; or if the Lt. Kernal is missing its host adapter, the Rear Admiral host adapter could be used in its place.

Also available for the Commodore 64 was the Creative Micro Designs CMD HD-Series. Much like the Commodore 1541 floppy drive, the CMD HD could connect to the Commodore 64's serial bus, and could operate independently of the computer with the help of its on-board hardware. A CMD HD series drive included its own SCSI controller to operate its hard drive mechanism, in addition to hosting a battery powered real-time clock module for the time-stamping of files. The stock operating speeds of the CMD HD-Series units were not very much faster than the stock speeds of a 1541 floppy drive, but the units were fully JiffyDOS compatible. Faster parallel transfers were possible with the addition of another CMD product, the CMD RAMLink and a special parallel transfer cable. With this arrangement, the performance of the system doubled that of the Lt. Kernal. One advantage the CMD products had was software compatibility, especially with GEOS, that prior solutions lacked. CMD ultimately missed opportunities to develop any features for the drive's auxiliary port (such as a printer spooler feature promised in the CMD HD user manual). Support for external SCSI devices (such as CD-ROM and Zip drives) was also noticeably missing. SCSI devices could be connected and chained to the external SCSI port, but could not be used from the HD without workarounds or special software.

The ICT DataChief included a 20MB hard drive, along with an Indus GT floppy drive, along with a 135-watt power supply in a case designed to house an IBM PC Compatible computer.[4]

User operation of these hard drive subsystems was similar to that of Commodore's floppy drives, with the inclusion of special DOS features to make best use of the drive's capabilities and to effectively manage the vast increase in storage capacity (up to a maximum of 4GB). An unavoidable problem was that total 1541 compatibility could not be achieved, which often prevented the use of copy-protected software, software fastloaders, or any software whose operation depended on exact 1541 emulation.

The enthusiast-built "IDE64 interface" was designed late in

the 1990s, attaching itself in the Commodore 64's expansion port, and allowing users to attach common IDE hard drives, CD-ROM and DVD drives, ZiP and LS-120 floppy drives to their Commodore 64s. Later revisions of the interface board provided an extra compact flash socket. The IDE interface's performance is comparable to the RAMLink in speed, but lacks the intelligence of SCSI. Its main advantage lies in being able to use inexpensive commodity hard drives instead of the more costly SCSI units. 1541 compatibility is not as good as commercially developed hard drive subsystems, but continues to improve with time.

In late 2011, MyTec Electronics developed and sold the Rear Admiral Thunderdrive, a clone of the CMD HD. Though using more modern components and a smaller form factor in comparison to the CMD HD, the Thunderdrive maintained full compatibility with the CMD HD.

2.8.2   Input/Output

Commodore MPS 802

Printers

A number of printers were released for the Commodore 64, both by Commodore themselves and by third-party manufacturers.

Commodore-specific printers were attached to the C64 via the serial port and were capable of being daisy chained to the system with other serial port devices such as floppy drives. By convention, printers were addressed as device #4-5 on the Commodore serial bus.

Dot-matrix   A series of dot-matrix printers were sold by Commodore, including the MPS 801 (OEM Seikosha GP 500 VC) and the MPS 803, although many other third-party printers like the Okimate 10 and Okidata 120 were popular too - some having more advanced printing features than any of Commodore's models. Most Commodore-branded printers were rebranded C. Itoh or Epson models with Commodore serial interface.

Daisy wheel Commodore also produced the DPS-1101 daisy wheel printer, which produced letter quality print similar to a typewriter, and which typically cost more than the computer and floppy disk drive together. The MPS-1000 dot matrix printer was introduced along with the C-128.[5] Commodore 1526 is a rebranded MPS 802.[6][7]

Commodore 1520 plotter

Plotter A mini plotter device, the Commodore 1520, could plot graphics and print text in four colors by using tiny ballpoint pens.

The 1520 was based upon the Alps Electric DPG1302, a mechanism which also formed the basis of numerous other inexpensive plotters for home computers of the time (e.g. the Atari 1020).[8][9]

Third-party printer interfaces and buffers Since there were severe shortcomings of early Commodore printers, CARDCO released the Card Print A (C/?A) printer interface that emulated Commodore printers by converting the Commodore-style IEEE-488 serial interface to a Centronics printer port to allow numerous 3rd-party printers to be connected to a Commodore 64, such as Epson, Okidata, C. Itoh.[10] A second model, a version that supported printer graphics was released called the Card Print +G (C/?+G), supported printing Commodore graphic characters using ESC/P escape codes. CARDCO released additional enhancements, including a model with RS-232 output, and shipped a total over 2 million printer interfaces. Xetec also released a series of printer interfaces. With a parallel interface, the QMS KISS laser printer, the most inexpensive then available at $1995, could be used.[11] Later,

CMD created the GeoCable which allowed PS2-type inkjet and laser printers to work under GEOS with a special device driver.

Printer buffer with 64 kB RAM for the IEC IEEE-488 serial bus existed too, like the "Brachman Associates Serial Box Print Buffer".[12]

Input devices

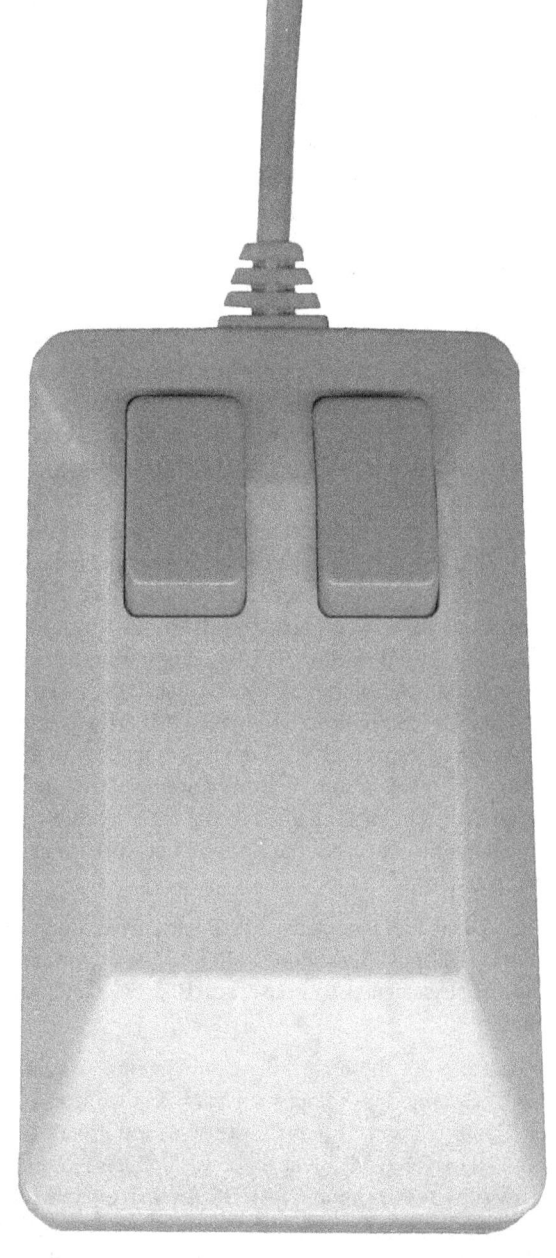

Commodore mouse

Commodore produced joystick controllers for the Com-

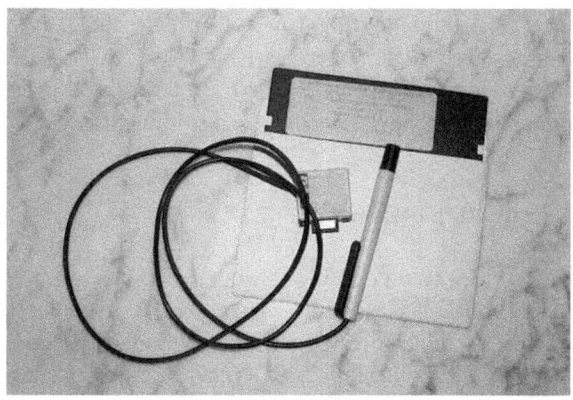

C64 Lightpen with its Software of the Company Rex-Datentechnik

modore 64, largely compatible with Atari joysticks, as well as paddles (which were not Atari compatible). Commodore's paddles were originally intended for the VIC-20, and few C64 games could take advantage of them. Commodore's joysticks were often derided because they were not particularly robust, especially for extreme gameplay. Many gaming enthusiasts preferred third-party joysticks, while some enthusiasts even built their own joysticks and controllers for the Commodore 64, or modified controllers from other systems to work on it. While the Commodore 64 only had two joystick ports for use, a few different kinds of joystick adapters were constructed by enthusiasts, which allowed up to four or eight joysticks to be used on the Commodore 64, with appropriate programming. Only about 20 games (by 2011) can take advantage of these however.

The "Atari CX85 Numerical Keypad" consists of a numeric keypad featuring the 17 keys [escape], [no], [delete], [yes], 0-9, [.], [-] and [+/enter].[13] It connects to the C64 joystick port using the Atari 2600 style interface with a DB9F plug.[14][15]

Commodore had two models of computer mouse, namely the 1350 and the 1351. These were used with GEOS as well as software such as Jane and Magic Desk. The earlier 1350 was only capable of emulating a digital joystick, by sending rapid 8 directional signals as it was moved, and thus was not very useful. The later 1351 used a more traditional proportional mode, sending signals to the computer that indicate amount and direction of movement. The 1351 also supported a mode identical to that of the 1350. CMD's SmartMouse was compatible with 1351-aware and also included a third button and a built in real-time clock module as well. The NEOS mouse also existed, but it was not compatible with 1351-aware software as it was simply a joystick emulator.

Several Companies produced Lightpens with its own drawing software for the Computer, e.g. the Inkwell light pen which was compatible with GEOS.

The Koala Pad graphics tablet was also available, came with its own paint software, and was compatible with GEOS as well. Suncom's Animation Station was another graphics tablet for the C64.[16]

Car positioning system

A senior test technicians at CGAD Productions operations developed and installed the CarPilot *Computerized Automotive Relative Performance Indicator and Location of Transit*. Which may be one of the first GPS type mapping systems to be tested, circa 1984. It utilizes a Commodore 64, 12V DC to 5V DC converter, video player/recorder, datasette, and a TV monitor.[17]

The monitor page 1 displays battery voltage, water temperature, engine oil pressure, fuel level, vehicle speed, engine rotation speed, lock/no-lock condition of the automatic transmission torque converter, and on/off condition of the air conditioning clutch. All except the last two were incorporated with a "buzzer" alarm system that indicate malfunction. Another feature is the one-second-precision 24-hour clock. Estimated arrival time with 1s precision, distance traveled which is incremented every 80 meters and estimated distance to arrival that is also decremented with same value, 80 meters.[17]

Page 2 displayed the vehicle position along the map. Vehicle location indication is calculated from distance traveled. The accuracy of the vehicle location is dependent of the digital map construction and the accuracy of the local map used to construct the digital map. The best hope for accuracy is 800 m. But accuracy of one car length in 35 km has been realized. The use assembly language was necessitated to keep up with sensor input. One advantage with the system is the ability to create one's own digital maps and thus eliminate the need to buy such ones for every trip. The software to accomplish this task was written in Basic.[17]

Robotics

With *computing*, *robot trainer*, and *plotter-scanner*, Fischertechnik rose as the first manufacturer of modular building blocks into the computer age. Interfaces for all popular home computers at the time were made, including Apple II, Commodore 64 and Acorn, and later for Schneider, Atari ST and IBM PC. Programming languages to drive the models included GW-BASIC, Turbo Pascal and in the later kits (1991) an in-house programming tool Lucky Logic.

The "Commocoffee 64" is an espresso maker controlled by the C64[18] in 1985.[19]

See also: List of educational programming lan-

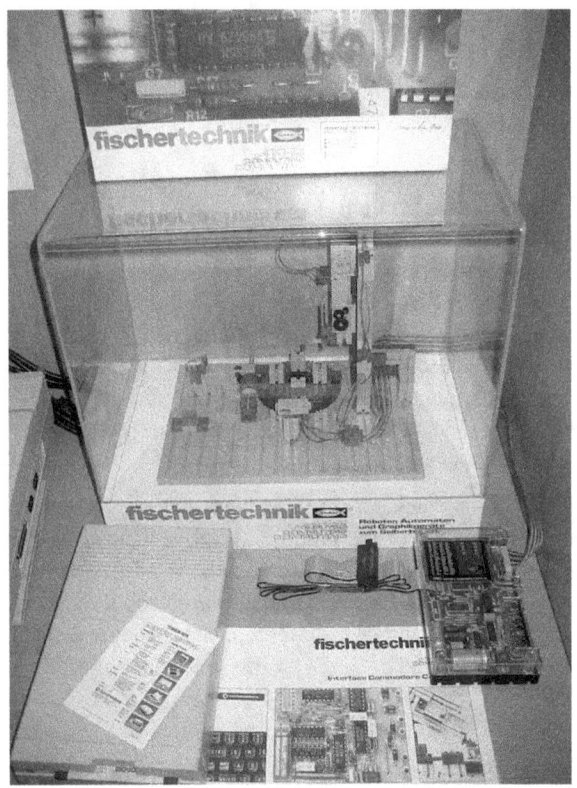

Fischertechnik computing *with a C64 interface*

guages

Relay controller

The Handic "VIC REL" controller provides protected input and output using 6 relay outputs and 2 optocoupler inputs. The output relays are capable of 24 V / 10 W and the inputs respond to 5-12 V DC. The device also provides (+5V) and (−5V) at 50 mA to activate inputs. The device is programmed on the VIC-20 with *POKE 37138,63* and I/O at *37136*. And on C64 with *POKE 56579,63* and I/O at *56577*. The intended applications were burglar alarms, garage doors, door locks, heating elements, lamps, transmitters, remote controllers, valves, pumps, telephones, accumulators, irrigation systems, electrical tools, stop watches, ventilators, humidifiers, etc.[20]

Analog to digital converters

There are audio Analog-to-digital converters (A/D) like the "A/D Wandler (DELA 87393)" based on 8-bit ADC0809 chip for the C64/128[21] with a maximum sampling frequency of 10 kHz.[22] and the Sound Ultimate Xpander 6400 (SUX 6400) based on the 8-bit ADC0804 chip with a maximum sampling frequency of 11 kHz. Plain sound digitizers like "Sound Digitizer (REX 9614)" that converts analog sound into 2-bit samples.[21] The latter could also be accomplished using the Datasette and software tricks.[23][24]

Biofeedback EEG/EMG

In 1987 there was a cartridge port device to measure EEG directly for use in exercise programs, called "BodyLink" produced by the company Bodylog in New York, USA.[25] Schippers-Medizintechnik in Germany produced a user port attached EMG device to allow a physician to analyze such things as stress level, and assisting in finding a better position for work.[26]

Handscanners

The "Scanntronik Handyscanner 64" is a hand held scanner that uses the C64 user port.[1][27]

Frame grabbers

Frame grabbers like the "PAL Colour Digitizer" that connect via the user port, will turn an analog composite video frame into a digital picture on the C64.[1] The "Print Technik Video Digitizer" connects via the user port and uses CVBS video signal that has to be still for 4 seconds in order to be sampled and can then be saved either as 320×200 monocolour or 160×200 multicolour (4 colours).[28]

Video generator

80 column mode could be used by installing the "BI-80" cartridge released 1984[29] from "Batteries Included" which is built around the 6545 video chip. It includes an expansion ROM that adds BASIC 4.0 commands. One can control which 40/80 column mode is active by software. On power up, the 40-column mode is active.[30][31]

Another 80 column card using the cartridge port was the "DATA20 XL80" introduced in 1984[32] Costing 400 000 Lira in 1985.[33]

The "Z80 Video Pack 80" enabled black and white 80-column screen and CP/M using a Zilog Z80.[15]

Teletext

To download pages and software transmitted via the teletext broadcast system. The UK company "Microtext" provided their "Teletext adaptor" and tuner that interfaced with the

TV-aerial and the C64/128 user port. Software was provided on a C-10 tape.[1][34] Which were priced at 114.80 GBP inc. p/p in 1987.[35]

2.8.3 Communication

Modems

Commodore VIC Modem

As Commodore offered a number of inexpensive modems for the C64, such as the 1650, 1660, 1670, the machine also helped popularize the use of modems for telecommunications.[36][37] The 1650 and 1660 were 300 Baud, and the 1670 was 1200 baud. The 1650 could only dial Pulse. The 1660 had no sound chip of its own to generate Touch Tones, so a cable from the monitor /audio out was required to be connected to the 1660 so it could use the C64 sound chip to generate Touch Tones. The 1670 used a modified set of Hayes AT commands.

This modem is required for Medical Manager for EDI operations.

The Commodore 1650 shipped with a rudimentary piece of terminal software called Common Sense. It provided basic Xmodem functionality and contained a 700 line scrollback feature.

In the United States, Commodore offered the Commodore Information Network, a CompuServe SIG devoted to its products and users. Later, Quantum Computer Services (which became America Online) offered an online service called Quantum Link for the C64 that featured chat, downloads, and online games. In the UK, Compunet was a very popular online service for C64 users (requiring special Compunet modems) from 1984 to the early 1990s. In Australia, Telecom (now Telstra) ran an online service called Viatel and sold modems for the C64 for use with the service. In Germany the very restrictive rules of the state-owned

telephone system prevented widespread use of inexpensive, non-telco licensed modems, prompting the use of inferior acoustic couplers instead. Access to Bildschirmtext, the state-owned telco's own dial-up online service, was possible via special add-on hardware like the Commodore "BTX Decoder Modul" [38] or the Commodore "BTX Decoder Modul II".[1][39][40]

Radio communication

"Microlog AIR-1 Radio Interface Cartridge" that use the cartridge port with builtin ROM software for RTTY and morse code communications.[41]

"RTTY-CW Interface C-64" uses the User port for RTTY communications.[42]

"Auerswald ACC-64" longwave time signal receiver.[43]

RS-232 port

Like the VIC-20, the C64 lacked a real UART chip such as the 6551 and used software emulation. This limited the maximum speed to an error-prone 2400 bit/s. Third-party cartridges with UART chips offered better performance.

Later in the Commodore 64's life, CMD developed two serial communications cartridges for Commodore Computers, the "Swiftlink" (1990[44] - 38 400 bit/s)[45] and the "Turbo 232" (1997[46] - 230 400 bit/s).[47] The latter was capable of handling a 56k Hayes modem reliably at full speed on a Commodore 64, enabling reasonable dial-up internet access speeds.

The Retro-Replay expansion cartridge enabled the addition of the **Silver Surfer** add-on serial board, which also enabled 56k modem connections, and the **RR-Net** add-on serial board, which allows for broadband internet access, as well as LAN.

Also, on November 5, 2005 Quantum Link Reloaded was launched enabling C64 enthusiasts to experience all the features of the original Quantum Link service in present-day with some enhancements for free.

IEEE-488

The Commodore 64 IEEE-488 Cartridges were made by various companies, but Commodore themselves never made one for the Commodore 64/128 family. One of uses were harddiscs like the Commodore D9060.

Some other interfaces without pictures available:

- E-LINK Serial to IEEE Interface. (contains 65C02, 6522 and 4 kB ROM)

- Buscard II Interface. (contains a 6532, 6821 (PIA) and 8 kB ROM, and a 256 byte PROM)

- INTERPOD - An Interface box, that converts IEEE-488 to CBM (IEC) serial & RS-232 serial.

2.8.4 Other peripherals

Commodore 1702 video monitor

The Commodore 1701 and 1702 were 13-inch (33 cm) color monitors for the C64 which accepted as input either composite video or separate chrominance and luminance signals, similar to the S-Video standard, for superior performance with the C64 (or other devices capable of outputting a separated signal). Other monitors available included the 1802 and 1902. Introduced in 1986, the 1802 featured separate chroma and luma signals, as well as a composite green screen mode suitable for the C-128's 80 column screen.[48] The 1902 had a true RGBI 80-column mode compatible with IBM PCs.

Early in the Commodore 64's life, Commodore released several niche hardware enhancements for sound manipulation. These included the "Sound Expander", "Sound Sampler", "Music Maker" overlay, and External music keyboard. The Sound Expander and Sound Sampler were both expansion cartridges, but had limited use. The Sound Sampler in particular could only record close to two seconds of audio, rendering it largely useless. The Music Maker was a plastic overlay for the Commodore 64 "breadbox" keyboard, which included plastic piano keys corresponding to keys on the keyboard. The External keyboard was an add-on which plugged into the Sound Expander. These hardware devices did not sell well, perhaps due to their cost, lack of adequate software, marketing as home consumer devices, and an end result that turned many serious musicians off.

Possibly the most complex C64 peripheral was the Mimic Systems Spartan, which added an entire new computer architecture to the C64, with its own 6502 CPU and expansion bus, for software and hardware compatibility with the Apple II series. Announced shortly after the Commodore 64 itself at a time when little software was available for the machine, the Spartan did not begin shipping until 1986, by which time the C64 had acquired an extensive software library of its own.[49] Essentially an Apple II+ compatible computer that used the 64's keyboard, video output, joysticks, and cassette recorder, the Spartan included 64kB RAM, a motherboard with a 6502 CPU on a card, 8 Apple-compatible expansion slots, an Apple-compatible disk controller card, and a DOS board to add to your 1541 disk drive. The DOS board was optional, but if it was not installed an Apple Disk II or compatible drive would be required to load software. The long delay between announcement and availability, along with heavy promotion including full-page ads running monthly in the Commodore press, made the Spartan an infamous example of vaporware.

Gamesware produced a gaming peripheral for the Commodore 64 in 1988, where a target board was attached to the computer using the RS-232 port to enable use of its *Gamma Strike* suite of games.

CMD produced a SID symphony cartridge later in the Commodore's life. A reworking of the original Dr. T's SID Symphony cartridge, this cartridge gave the Commodore another SID chip for use to play stereo SID music. This saved Commodore 64 users from needing to modify their computer motherboards to enable it with dual SID chips.

Creative Micro Designs (CMD) was the longest-running third-party hardware vendor for the Commodore 64 and Commodore 128, hailed by some enthusiasts as being better at supporting the Commodore 64 than Commodore themselves. Their first commercial product for the C64 was a KERNAL based fast loader and utility chip called JiffyDOS. It was not the first KERNAL-based enhancement for the C64 (SpeedDOS and DolphinDOS also existed), but was perhaps the best implemented. The benefits of a KERNAL upgrade meant that the cartridge port was free for use (which would have normally been taken up by an Epyx FastLoad cartridge or an Action Replay), however the downside meant that one had to manually remove computer chips from the C64's motherboard and associated floppy drives to install it. Aside from the usual 1541 fast load routines, JiffyDOS contained an easy to use DOS and a few other useful utilities.

RAM expansions

Over the years, a number of RAM expansion cartridges were developed for the Commodore 64 and 128. Commodore officially produced several models of RAM expansion cartridges, referred to collectively as the 17xx-series Commodore REUs. While these devices came in 128, 256, or 512 kB sizes, third-party modifications were quickly developed that could extend these devices to 2 MB, although some such modifications could be unstable. Some companies also offered services to professionally upgrade these devices.

Typically, most Commodore 64 users did not require a RAM expansion. Very little of the available software was programmed to make use of expansion memory. The cost of the units (and the requirement to add a heavy-duty power supply) also was a factor in the limited usage of RAM expansion cartridges. The volatility of DRAM was also a factor in the limited usage, as the RAM expansion cartridges were normally used for fast RAM disk storage, data stored on them would be lost at any power failure.

Aside from power-supply problems, the other main downfall of the RAM expansions were their limited usability due to their technical implementation. The RAM in the expansion cartridges was only accessible via a handful of hardware registers, rather than being CPU-addressable memory. This meant that users could not access this RAM without complicated programming techniques. Furthermore, simply adding the RAM expansion did not provide any kind of on-board RAM disk functionality (though a utility disk was supplied with some REUs, which provided a loadable RAM disk driver).

One popular exception to the disuse of the REUs was GEOS. As GEOS made heavy use of a primitive, software-controlled form of swap space, it tended to be slow when used exclusively with floppy disks or hard drives. With the addition of an REU, along with a small software driver, GEOS would use the expanded memory in place of its usual swap space, increasing GEOS' operating speed.

Due to the lack of available 17xx-series Commodore REUs, and then their later discontinuation, Berkeley Softworks, the publishers of GEOS, developed their own 512 kB RAM expansion cartridge - the GeoRAM. This device was purposely designed for use with GEOS, although some REU-aware programs were later adapted to be able to use it. Some time later, the GeoRAM was cloned by another company to form the BBGRAM device (which also sported a battery backup unit). The GeoRAM used a banked-memory design where portions of the external SRAM were banked into the Commodore 64's CPU address space. This method provided substantially slower transfer speeds than the single-cycle-per-byte transfer speeds of the Com-

modore REUs. A benefit of using SRAMs was lower power consumption which did not require upgrading the Commodore 64's power supply.

Eventually the Super 1750 Clone, a third-party clone of Commodore's RAM expansions was developed, designed in such a way as to eliminate the need for a heavy-duty power supply.

PPI devised their own externally powered 1 or 2 MB RAM expansion, marketed as the PPI/CMD RAMDrive, which was explicitly designed to be used as a RAM disk. Its primary feature was that the external power supply kept the formatting and contents of the RAM safe and valid while the computer was turned off, in addition to powering the device in any case. A driver was provided on the included utilities disk to allow GEOS to use the RAMdrive as a regular 'disk' drive.

CMD later followed up with the RAMLink. This device operated similar to the RAMDrive, but could address up to 16 MB of RAM in the form of a 17xx-series REU, Geo-RAM, and/or an internal memory card, which also provided a battery-backed realtime clock for file time/date stamping of files saved to it. It also features a battery backup, thus preserving the RAM's contents. Drivers were provided with the RAMLink to allow GEOS to use its memory as either a replacement for swap space, or as a regular 'disk' drive.

CMD's Super CPU Accelerator came after this, and could house up to 16 MB of direct, CPU-addressable RAM. Unfortunately, there was no on-board or disk-based RAM disk functionality offered, nor could any existing software make use of the directly addressable nature of the RAM. The exception is that drivers were included with the unit to explicitly allow GEOS to use that RAM as a replacement for swap space, or as a regular 'disk' drive, as well as to make use of the acceleration offered by the unit.

EPROM programmers

Programmers for EPROMs like 2716 - 27256 using common programming voltages (Vpp) of 12.5, 21, and 25 V were available by connecting a device to the user port of the C64.[50] These devices could cost 100 USD in 1985. The device often included a zero insertion force (ZIF) socket and a LED indicating when the EPROM chip was being programmed.[51] The cartridge port was also used by some programmer devices.[21]

Freezer, Reset, and Utility cartridges

Probably the most well-known hacker and development tools for the Commodore 64 included "Reset" and "Freezer" cartridges. As the C64 had no built-in soft re-

Micro Maxi Prommer, EPROM burner for C64 user port

set switch, reset cartridges were popular for entering game "POKEs" (codes which changed parts of a game's code in order to cheat) from popular Commodore computer magazines. Freezer cartridges had the capability to not only manually reset the machine, but also to dump the contents of the computer's memory and send the output to disk or tape. In addition, these cartridges had tools for editing game sprites, machine language monitors, floppy fast loaders, and other development tools. Freezer cartridges were not without controversy however. Despite containing many powerful tools for the programmer, they were also accused of aiding software pirates to defeat software copy protections. Perhaps the best known freezer cartridges were the Datel "Action Replay", Evesham Micros Freeze Frame MK III B, Trilogic "Expert", "The Final Cartridge III", and Super Snapshot cartridges.

The Lt. Kernal hard drive subsystem included a push button on the host adapter called ICQUB (pronounced "ice cube"), which could be used to halt a running program and capture a RAM image to disk. This would work with most copy-protected software that did not do disk overlays and/or bypass the KERNAL ROM jump table. The RAM image was runnable only on the Lt. Kernal system on which it was captured, thus preventing the process from being used to pirate software.

Music and Synthesizer utilities

As the Commodore 64 featured a digitally controlled semi-analogue synthesizer as its sound processor, it was not surprising to discover an abundance of software and hardware designed to expand upon its capabilities.

Various assemblers, notators, sequencers, MIDI editing and mixer automation software were created which allowed users and programmers to create or record musical pieces of impressive technical complexity. Some software of note has included the Kawasaki Synthesizer range, Music System notation and MIDI suite, the MIDI-compatible Instant Music 'idiot-proof' sequential composer, and the Steinberg Pro-16 MIDI sequencer, the precursor to Cubase.

Notable hardware included various brands of MIDI cartridges, plug-in keyboards (such as the Color Tone or the Sound Chaser 64), Commodore's own SFX range which included a sound sampler and Sound Expander plug-in synthesizer and keyboard, the more recent Commodulator oscillator wheel and the Prophet 64 sequencer and synthesizer utility cartridge. The Passport Designs MIDI Interface is said to be one of the best designs and had the most software supported model available.[15]

Recently a few professional musicians have used the Commodore 64's unique sound to provide some or all of the synthesizer parts required for their performances or recordings; an example being the band Instant Remedy. Also noteworthy is the Commodore 64 Orchestra who specialize in rearranging and performing music originally composed and coded for the Commodore 64 games market. Its patron is celebrated Commodore composer Rob Hubbard.

Apple II+ emulation box

The Mimic Systems "Mimic Spartan Apple II+ compatibility box" enabled C64 users to run Apple II+ software.[52] It came with the "DOS Card" addition, an Apple II disk controller that was installed inside the Commodore 1541 disk drive, between the floppy logic board and the drive mechanism. In normal mode the circuit simply passed signals through but at the flick of a switch it could take over the mechanism and turn the drive into an Apple II drive. The potential for grave damage to both Apple II and 1541 floppies was enormous and often happened. The box had 24 jumpers to configure. Applesoft BASIC was included and very compatible, since it was created by disassembling the binary from the Applesoft ROM and reordering the assembly level instructions such that the binary image would be different. One could set up various debugging and use slave computing to enable fast 3D rendering etc. The box had functionality to switch video between C64 and Apple. The second advertisement were put into the COMPUTE!'s Gazette in 1986.[53]

CP/M with Z80 CPU cartridge

The Commodore C64 CP/M Cartridge used the C1541 floppy drive that was incapable to read any existing CP/M disk format. The cartridge were equipped with a Zilog Z80 CPU running at circa 3 MHz.[54] On the C128 the INT and IORQ signals are used such that the Z80 can make use of interrupts.[55]

Present and Future devices

While CMD no longer produces Commodore hardware, new peripherals are still being developed and produced, mostly for mass storage or networking purposes.

CPU accelerators

Like the Apple II family, third-party acceleration units providing a faster CPU appeared late in the C64's life. Due to timing issues with the VIC-II chip - the same issues that caused the 1540 disk drive to be incompatible and the 128's "fast mode" to be 80 column-only - CPU accelerators for the 64 were much more complex and expensive to implement than for other computers. So while accelerators based on the WDC 65C02, usually running at 4 MHz, and on the 65816 at up to 20 MHz appeared, they appeared too late and were too expensive to gain widespread use.

The first CPU accelerator seen was called the "Turbo Process" by a Bonn, Germany, based company called Roßmöller GmbH. It used a Western Design Center 65816 running at 4.09 MHz. Code ran from faster static RAM on the accelerator expansion port cartridge. As the VIC chip can only see the internal DRAM memory, writes had to be mirrored to the internal memory, write cycles would slow the operation of the processor to accomplish this.

The *Turbo Master CPU*, produced by US based Schnedler Systems, was a blue expansion port device which clocked in at 4.09 MHz. It also had a JiffyDOS option. It was a copy of the Turbo Process system. Early Turbo Process circuit boards shipped with PAL chips that did not have their security fuses blown, this made copying the design quite easy. The Turbo Master CPU had one beneficial modification, the bit to toggle the high-speed mode on was "0" in memory location $00 as opposed to the "1" the Turbo Process. A lot of software would write zeros to this location turning off the high-speed mode on the Turbo Process - this was considered a design flaw that was fixed by the Turbo Master. No known litigation took place over the copying of the German company's design.

The most well-known accelerator for the C64 is probably Creative Micro Designs' SuperCPU, which gives the C64 a 20 MHz processor (instead of ~1 MHz) and up to 16 MB of RAM if combined with CMD's *SuperRam-Card*. Understandably, due to a very limited "market" and number of developers, there has not been much software tailored for the SuperCPU to date— however GEOS was supported. Among the few offerings available include the GEOS-compatible operating system, Wheels; a Wheels-based web browser called "The Wave", a Unix/QNX-like graphical OS called Wings, some demos, various classic games modified for use with the SuperCPU, and a shooter game in the old *Katakis*-style called *Metal Dust*.

The MMC64 cartridge allows the C64 to access MMC- and SD flash memory cards. And several revisions and add-ons have been developed for it to take advantage of extra features. It features an Amiga clock port for connecting a RR-Net Ethernet-Interface, an MP3 player add-on called 'mp3@c64' has even been produced for it.

In February 2008, Individual Computers started shipping the MMC Replay. It unites the MMC64 and the Retro Replay in one cartridge, finally built with proper case-fit in mind (even including the RRnet2 Ethernet add-on). It contains many improvements, such as C128 compatibility, a built-in .d64 mounter (not speedloader-compatible though, because the 1541 CPU is not emulated), 512 kB ROM for a total of eight cartridges, 512 kB RAM, a built-in flash-tool for cartridge images and wider support for various types of cartridges (not merely Action-replay-based).

In April 2008, the first batch of *1541 Ultimate* shipped, a project by the hobbyist VHDL-developer Gideon Zweijtzer. This is a cartridge that carries an Action Replay and Final Cartridge (whatever the user prefers) and a very compatible FPGA-emulated 1541 drive that is fed from a built-in SD-card slot (.d64, prg etc.). The difference to other SD-based and .d64 mounting cartridges like the MMC64, Super Snapshot 2007 or MMC Replay is, that the 6502 that powers the 1541 Floppy and the 1541's mechanical behavior (even sound) is fully emulated, making it theoretically compatible with almost anything. Fileselection and management is done via a third button on the cartridge that brings up a new menu on screen. The 1541 Ultimate also works in standalone mode without a c-64, functioning just like a normal Commodore 1541 would. Disk-selection of .d64s is then done via buttons on the cartridge, power is supplied via USB. There is a "Plus-Version" available with an extra 32 Megabytes of RAM (as REU and for future use), the basic version has just enough RAM for the advertised functions to work. In October 2008, the second and third batch of 1541 Ultimates were produced to match the public demand for the device. The regular version without the 32MB RAM was dropped since there was no demand for it. Due to public demand there is also a version with Ethernet now. In 2010 a completely new PCB and software has been developed by Gideon Zweijtzer to facilitate the brand new 1541-Ultimate-II cartridge.

The IDE64 interface cartridge provides access to parallel ATA drives like hard disks, CD/DVD drives, LS-120, Zip drives, and CompactFlash cards. It also supports network drives (PCLink) to directly access a host system over various connection methods including X1541, RS-232, Ethernet and USB. The operating system called IDEDOS provides CBM/CMD compatible interface to programs on all devices. The main filesystem is called CFS, but there's read-only support for ISO 9660 and FAT12/16/32. Additional features include BASIC extension, DOS Wedge, file manager, machine code monitor, fast loader, BIOS setup screen.

Today's computer mice can be attached via the Micromys interface that can process even optical mice and similar. There are also various interfaces for plugging the 64 to a PC keyboard.

A special board for converting Commodore 64 video signals to standard VGA monitor output is also currently under development. Also a board to convert the Commodore 128's 80 column RGBI CGA-compatible video signal to VGA format was developed in late 2011. The board, named the C128 Video DAC, had a limited production run and was used in conjunction with the more widespread GBS-8220 board.

In September 2008, Individual Computers announced the Chameleon, a Cartridge for the Expansion Port that adds a lot of previously unseen functionality. It has a Retro-Replay compatible Freezer and MMC/SD-Slot, 16 MB REU and a PS/2 connector for a PC Keyboard. Support for a network adapter and battery-backed real time clock exists. The cartridge does not even have to be plugged into a Commodore 64 and can be used as a standalone device using USB power. Since the cartridge essentially also includes a Commodore One it is possible to include a VGA Port that outputs the picture to a standard PC monitor. The Commodore One core also allows the cartridge to be used as a CPU accelerator, and a core to run a Commodore Amiga environment in standalone mode also exists. Unlike most other modern day C64 hardware, this cartridge actually ships with a bright yellow case. Shipping was announced for Q1/2009, and currently the cartridge is available, although the firmware is in a beta state. A standalone mode docking station is under development.

Retro Innovations is shipping the *uIEC*[56] device, which utilizes the core design of the *SD2IEC* project to provide a mass media solution for Commodore 8-bit systems that utilize the Commodore IEC Serial Bus. NKCElectronics of Florida is shipping SD2IEC hardware which uses the sd2iec firmware. Manosoft sells the C64SD Infinity, another SD card media solution which uses the sd2iec firmware.

In Summer of 2013, another commercial variant of the SD2IEC-Device appears on market, the *SD2IEC-evo2* from 16xEight.[57] This device uses an bigger uC (ATmega1284P) and has some extras such as Battery backed-up RTC, connector for LC-Display, Multicolour Status-LED, and so on already on board.

2.8.5 Notes

1. ^ Many users came to dread the telltale "RAT-AT-AT-AT-AT" knocking noise, since such knocking contributed to eventual disk drive alignment failure.

2. ^ A modification could be made to older model Commodore 64 motherboards to piggy-back a secondary SID sound chip to the original SID chip. The resulting modification enabled the Commodore 64 to play sound in 6-channel stereo with the appropriate software.

3. ^ The Commodore 64 had documented cartridge port pins which could be crossed to achieve a reset. In an attempt to activate game "reset" and various cheats, a large number of Commodore 64 users attempted to reset their machines by manually touching these pins 1 and 3 with wire while the computer was switched on. Many users made mistakes and missed the correct pins, blowing their C64's fuse and resulting in a costly repair. This achievement was later known as the "Hamster Reset" in "Commodore Format" magazine. Some users soldered these pins to a button, which they mounted in the C64's case for handy resetting. Some programs utilized reset protection (by having the string 'CBM80' [58] at $8000 in the memory) which could be worked around by shorting pins 1-3-9 the same way as the "Hamster Reset" pin 9 (on the top side as opposed to pins 1 & 3 on the bottom) being the EXROM ROM expansion pin (thus overwriting data at $8000–$9fff).

2.8.6 See also

- Computers: Commodore 64, VIC-20

- Floppy Drives: Commodore 1541, 1551, 1570, 1571, 1581

- Commodore 64 disk / tape emulation

2.8.7 References

[1] "Hardware". bithunter.siz.hu. 2012-01-30. Retrieved 2013-06-21.

[2] "coll_quick_data_drive.jpg". bithunter.siz.hu. 2012-01-23. Retrieved 2013-06-21.

[3] "tt". web.tiscali.it. 2012-09-22. Retrieved 2013-06-17.

[4] "RUN Magazine issue 40".

[5] "Run Issue 30 Jun 1986".

[6] "Chronology of Commodore Computer History, Jack Tramiel". 090505 commodore.ca

[7] "Here be Commodore Computers. Be in Awe.". 090505 zimmers.net

[8] "What are the Atari 1020, 1025, 1027, and 1029 Printers?". *faqs.org (Atari 8-Bit Computers: Frequently Asked Questions section)*. Retrieved 2015-03-22. = Commodore 1520 / Oric MCP40 / Tandy/Radio Shack CGP-115 /..; made by ALPS [..] 20, 40 and 80-column modes

[9] "The Texas Instruments HX-1000 Printer/Plotter Photos". *Hexbus.com*. Other printer plotters that use variants of the ALPS DPG1302 plotter mechanism include the: Commodore 1520, Tandy CGP-115, Sharp CE-150, Atari 1020, Mattel Aquarius 4615

[10] "CARDCO Card Print A (C/?A) - Printer Interface For The Commodore 64 and VIC-20". *COMPUTE Magazine* (34): 251. March 1983.

[11] "RUN Magazine issue 36".

[12] "commodore.ca | Rare Commodore Computer Hardware Picture / Photo Gallery". commodore.ca. 2012-12-11. Retrieved 2013-06-21.

[13] "Commodore%2064_128%20Key%20Pad_Atari.jpg". commodore.ca. 2011-03-29. Retrieved 2013-06-21.

[14] "Review: Atari   CX85   Numerical   Keypad". atarimagazines.com.   May 1983.   Retrieved 2013-06-21.

[15] "Products | Commodore 64 History, Manuals & Photo's 64C 64GS". commodore.ca. 2011-03-30. Retrieved 2013-06-21.

[16] Infoworld Media Group, Inc (1984-07-09). *Software for the Suncom Graphics Tablet.*

[17] "commodore-64-car-pilot.jpg". commodore.ca. 2011-03-30. Retrieved 2013-06-21.

[18] "The Commocoffee-64 » Coolest Gadgets". coolest-gadgets.com. Retrieved 2013-06-21.

[19] "commocoffee-commodore-64-coffee-maker.jpg". commodore.ca. 2011-03-29. Retrieved 2013-06-21.

[20] "VIC REL" (PDF). bombjack.org. 2009-11-14. Retrieved 2013-06-21.

[21] "- Rex Datentechnik - Retroport". retroport.de. 2013-06-14. Retrieved 2013-06-21.

[22] "ADC0808/ADC0809 8-Bit µP Compatible A/D Converters with 8-Channel Multiplexer" (PDF). learn-c.com. 2010-04-15. Retrieved 2013-06-21.

[23] "Could the Datasette players play music cassette tapes too? - Commodore 64 (C64) Forum". lemon64.com. Retrieved 2013-06-21.

[24] "C64 Tape player - Commodore 64 (C64) Forum". lemon64.com. Retrieved 2013-06-21. 5 poke53265,0 10 for i=0 to 25:read a:poke49152+i,a:next:sys49152 90 data 120,165,1,41,223,133,1,162,0,160,15,169,16 91 data 44,13,220,240,251,142,24,212,140,24,212,208,243

[25] COMPUTE!'s GAZETTE, January 1987, Issue 43, Vol. 5, No. 1 |page=10

[26] "The C64 as a medical aid". mos6502.com. 2012-09-21. Retrieved 2013-07-06.

[27] "coll_handyscanner.jpg".   bithunter.siz.hu.   2012-01-23. Retrieved 2013-06-21.

[28] "coll_pal.jpg". bithunter.siz.hu. 2012-01-23. Retrieved 2013-06-21.

[29] "B I - 8 0 80 Column Display by Batteries Included" (PDF). mikenaberezny.com. Retrieved 2013-06-17.

[30] "BI-80 Display Adapter". mikenaberezny.com. 2012-01-28. Retrieved 2013-06-17.

[31] "coll_bi-80.jpg". bithunter.siz.hu. 2012-01-23. Retrieved 2013-06-21.

[32] "B80.jpg". web.tiscali.it. 2012-09-16. Retrieved 2013-06-17.

[33] "Data 20 Corporation XL 80 video a 80 colonne per C 64" (PDF). digitanto.it. 2010-02-13. Retrieved 2013-06-17.

[34] "coll_microtext.jpg". bithunter.siz.hu. 2012-01-23. Retrieved 2013-06-21.

[35] Your Commodore, Issue 35, August 1987, page 7

[36] http://www.zimmers.net/cbmpics/ouser1.html

[37] http://archive.org/stream/VIC-1600_VICMODEM_1982_Commodore/VIC-1600_VICMODEM_1982_Commodore_djvu.txt

[38] "- Hardware B-C - Retroport". retroport.de. 2013-06-14. Retrieved 2013-06-21.

[39] "coll_btx.jpg". bithunter.siz.hu. 2012-01-23. Retrieved 2013-06-21.

[40] "Bildschirmtext-Museum: Hardware-Btx-Decoder: Meine Sammlung". btxmuseum.de. Retrieved 2013-06-21.

[41] "coll_microlog_air-1.jpg". bithunter.siz.hu. 2012-01-23. Retrieved 2013-06-21.

[42] "empty". bithunter.siz.hu. 2012-01-23. Retrieved 2013-06-21.

[43] "coll_acc64.jpg". bithunter.siz.hu. 2012-01-23. Retrieved 2013-06-21.

[44] "Mike Naberezny – CMD SwiftLink RS-232". mikenaberezny.com. 2012-01-28. Retrieved 2013-06-17.

[45] "USR Modem - comp.sys.cbm | Google Groups". groups.google.com. 1996-08-06. Retrieved 2013-06-17.

[46] "File:Turbo232 top.jpg - ReplayResources". ar.c64.org. 2010-06-15. Retrieved 2013-06-17.

[47] "CMD Turbo232 High speed modem interface" (PDF). ar.c64.org. 2010-06-15. Retrieved 2013-06-17.

[48] "Commodore 1802 User's Manual".

[49] "RUN Magazine issue 36 December 1986".

[50] "empty" (PDF). bombjack.org. 2009-01-26. Retrieved 2013-06-21.

[51] "EPROM Programmers handbook for the C64 and C128" (PDF). bombjack.org. 2009-01-02. Retrieved 2013-06-21.

[52] "VC&G | [Retro Scan of the Week] Apple II Box for C64". vintagecomputing.com. 2013-03-25. Retrieved 2013-06-21.

[53] "Mimic Systems' Spartan | Applefritter". applefritter.com. 2013-06-21. Retrieved 2013-06-21.

[54] "Commodore 64 CP/M Cartridge". devili.iki.fi. 2006-02-24. Retrieved 2013-06-21.

[55] "Ruud's Commodore Site: C/PM-cartridge for the C64". baltissen.org. 2009-07-30. Retrieved 2013-06-21.

[56] Retro Innovations - uIEC

[57] SD2IEC-evo2

[58] The string 'CBM80' being represented by the hex bytes C3 C2 CD 38 30

2.8.8 External links

- Individual Computers - Makers of MMC64 and RR-series products

- 16xEight Digital Retrovation - Makers of innovative new hardware for Commodore 8-Bit Computers

- Protovision - Makers of various new hardware upgrades

- Lemon64 - Includes some of the best Commodore 64 music software

- Home Recording - Music discussion board thread linking to many others relevant to C64 music

- RUN Magazine Issue 39 May, 1986 special printer issue

- elektronik.si: Vic-Rel internal PCB

Manuals

Commodore

- Commodore VIC-1541 Floppy Drive: User Manual, Technical Reference

- Commodore VIC-1515 Printer: User Manual

- Commodore VIC-1525 Printer: User Manual

CARDCO

- CARDCO Card Print A (C/?A) Printer Interface: User Manual, Addendum

- CARDCO Card Print +G (C/?+G) Printer Interface: User Manual, Supplement

2.9 Commodore 1541

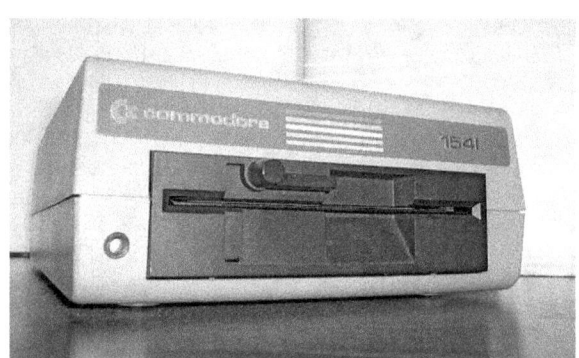

Front view of the second-most common version of the Commodore 1541 disk drive, with open disk slot: this version uses a Newtronics drive mechanism, and the rotating lever is used to engage the drive mechanism with the disk (i.e. to engage the hub clamp and load the disk heads) and to prevent removal of the disk while the mechanism is mechanically engaged.

The **Commodore 1541** (also known as the **CBM 1541** and **VIC-1541**) is a floppy disk drive (FDD) which was made by Commodore International for the Commodore 64 (C64), Commodore's most popular home computer. The best-known FDD for the C64, the 1541 is a single-sided 170-kilobyte drive for 5¼" disks. The 1541 directly followed the Commodore 1540 (meant for the VIC-20).

The disk drive uses group code recording (GCR) and contains a MOS Technology 6502 microprocessor, doubling as a disk controller and on-board disk operating system processor. The number of sectors per track varies from 17 to 21 (an early implementation of zone bit recording). The drive's built-in disk operating system is CBM DOS 2.6.

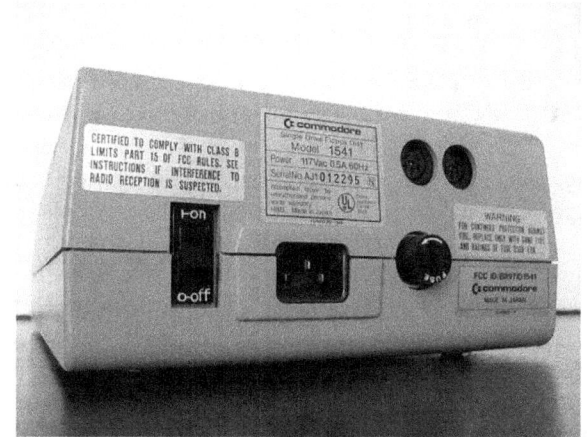

Back panel view of the Commodore 1541 disk drive

2.9.1 History

Introduction

The 1541 was priced at under US\$400 at its introduction. A C64 plus a 1541 cost about \$900, while an Apple II with no disk drive cost \$1295. The first 1541 drives produced in 1982 have a label on the front reading VIC-1541 and have an off-white case to match the VIC-20. In 1983, the 1541 was switched to having the familiar beige case and a front label reading simply "1541" along with rainbow stripes to match the Commodore 64.

By 1983, after a brutal home-computer price war that Commodore began, the C64 and 1541 cost under \$500. The drive became very popular, and became difficult to find. The company claimed that the shortage occurred because 90% of C64 owners bought the 1541 compared to its 30% expectation, but the press reported what *Creative Computing* described as "an absolutely alarming return rate" because of defects. The magazine reported in March 1984 that it received three defective drives in two weeks,[1] and the lead editorial in the December 1983 issue of *Compute!'s Gazette* reported that four of the seven drives the magazine had in its editorial offices had failed.

The early (1982–83) 1541s have a spring-eject mechanism (Alps drive), and the disks often fail to release. This style of drive has the popular nickname "Toaster Drive", because it requires the use of a knife or other hard thin object to pry out the stuck media just like a piece of toast stuck in an actual toaster (though this is inadvisable with actual toasters). This was fixed later when Commodore changed the vendor of the drive mechanism (Mitsumi) and adopted the flip-lever Newtronics mechanism, greatly improving reliability. In addition, Commodore made the drive's controller board smaller and reduced its chip count compared to the early 1541s (which had a large PCB running the length of the

case, with dozens of TTL chips). The beige-case Newtronics 1541 was produced from 1984-86.

Versions and third-party clones

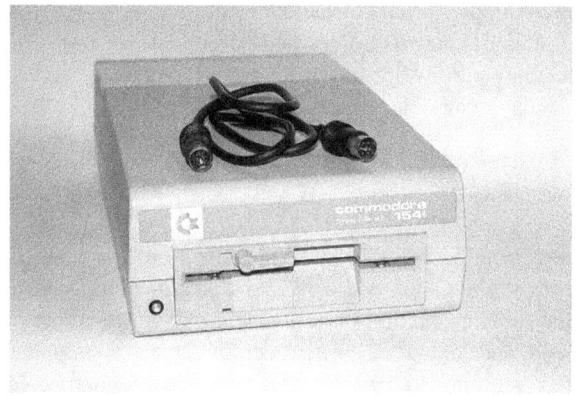

1541C, the first upgrade version

All but the very earliest non-II model 1541s can use either the Alps or Newtronics mechanism. Visually, the first models, of the *-VIC-1541* denomination, have an off-white color like the VIC-20 and VIC-1540. Then, to match the look of the C64, CBM changed the drive's color to brown-beige and the name to *Commodore 1541*.

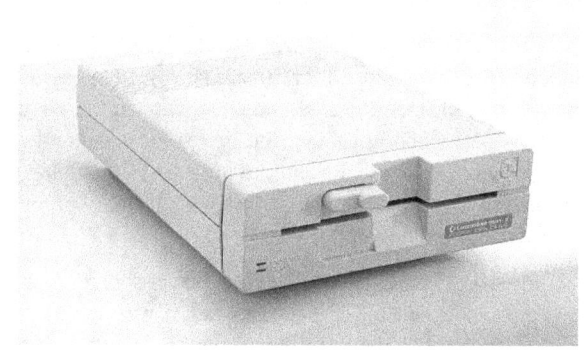

Commodore 1541-II, the second of two upgraded versions of the CBM 1541. The 1541-II has the more modern "radial handle" locking mechanism.

The 1541's numerous shortcomings opened a market for a number of third-party clones of the disk drive, a situation that continued for the lifetime of the C64. Well-known clones are the *Oceanic OC-118* a.k.a. *Excelerator+*, the MSD Super Disk single and dual drives, the *Enhancer 2000*, the *Indus GT*, and *CMD*'s *FD-2000* and *FD-4000*. Nevertheless, the 1541 became the first disk drive to see widespread use in the home and Commodore sold millions of the units.

In 1986, Commodore released the 1541C, a revised version that offered quieter and slightly more reliable operation and a light beige case matching the color scheme of the Commodore 64C. It was replaced in 1988 by the 1541-II, which uses an external power supply to provide cooler operation and allows the drive to have a smaller desktop footprint (the power supply "brick" being placed elsewhere, typically on the floor). Later ROM revisions fixed assorted problems, including a software bug that made the save-and-replace command unusable.

Successors

The Commodore 1570 is an upgrade from the 1541 for use with the Commodore 128, available in Europe. It offers MFM capability for accessing CP/M disks, improved speed, and somewhat quieter operation, but was only manufactured until Commodore got its production lines going with the double-sided 1571. Finally, the small, external-power-supply-based, MFM-based Commodore 1581 3½" drive was made, giving 800 KB access to the C128 and C64. By this time, however, many CBM users had shifted their attention to the 16/32-bit Amiga, and the 1581 was mostly sold to remaining GEOS users.

2.9.2 Design

Hardware

The 1541 does not have dip switches to change the device number. If a user added more than one drive to a system the user had to open the case and cut a trace in the circuit board to permanently change the drive's device number, or hand-wire an external switch to allow it to be changed externally.[2] It was also possible to change the number temporarily from the operating system.

The pre-II 1541s also have an internal power source, which generate much heat. The heat generation was a frequent source of humour. For example, *Compute!* stated in 1988 that "Commodore 64s used to be a favorite with amateur and professional chefs since they could compute and cook on top of their 1500-series disk drives at the same time".[3] A series of humorous tips in *MikroBitti* in 1989 said "When programming late, coffee and kebab keep nicely warm on top of the 1541." The *MikroBitti* review of the 1541-II said that its external power source "should end the jokes about toasters".

The drive-head mechanism installed in the early production years is notoriously easy to misalign. The most common cause of the 1541's drive head knocking and subsequent misalignment is copy-protection schemes on commercial software.[4] The main cause of the problem is that the disk drive itself does not feature any means of detecting when the read/write head reaches track zero. Accordingly, when a disk is formatted or a disk error occurs, the unit tries to move the head 40 times in the direction of track zero (although the 1541 DOS only uses 35 tracks, the drive mechanism itself is a 40-track unit, so this ensured track zero would be reached no matter where the head was before). Once track zero is reached, every further attempt to move the head in that direction would cause it to be rammed against a solid stop: for example, if the head happened to be on track 18 (where the directory is located) before this procedure, the head would be actually moved 18 times, and then rammed against the stop 22 times. This ramming gives the characteristic "machine gun" noise and sooner or later throws the head out of alignment.

The earlier 1541s are so unreliable that *Info* magazine joked, "Sometimes it seems as if one of the original design specs ... must have said 'Mean time between failure: 10 accesses.'". Users can realign the drive themselves with a software program and a calibration disk. What the user would do is remove the drive from its case and then loosen the screws holding the stepper motor that moved the head, then with the calibration disk in the drive gently turn the stepper motor back and forth until the program shows a good alignment. The screws are then tightened and the drive is put back into its case.[4]

A third-party fix for the 1541 appeared where the solid head stop was replaced by a sprung stop, giving the head a much easier life.[4] The later 1571 drive (which is 1541-compatible) incorporates track-zero detection by photo-interrupter and is thus immune to the problem. Also, a software solution, which resides in the drive controller's ROM, prevents the rereads from occurring, though this could cause problems when genuine errors did occur.

Interface

The 1541 uses a proprietary bit-serial derivative of the standardized IEEE-488 parallel interface, which is used on Commodore's earlier drives for the PET/CBM range of personal/business computers. To ensure a ready supply of inexpensive cabling for its home computer peripherals, Commodore chose standard DIN connectors for the serial interface. Disk drives and other peripherals such as printers are connected to the computer via a daisy chain scheme, necessitating only a single connector on the computer itself.

2.9.3 Throughput and software

Initially, Commodore intended to use a hardware shift register (one component of the 6522 VIA) to maintain relatively brisk drive speeds with the new serial interface. However, a hardware bug with this chip prevented the initial design from working as anticipated, and the ROM code was hastily rewritten to handle the entire operation in software. According to Jim Butterfield, this causes a speed reduction by a factor of five.[5]

As implemented on the VIC-20 and Commodore 64, CBM DOS transfers only about 300 bytes per second - compare the 300-baud data rate of the Commodore Datasette storage system - which translates to about 20 minutes to copy one disk—10 minutes of reading time, and 10 minutes of writing time. However, since both the computer and the drive can easily be reprogrammed, third parties quickly wrote more efficient firmware that would speed up drive operations drastically. Without hardware modifications, some "fast loader" utilities managed to achieve speeds of up to 4 kB/s. The most common of these products are the Epyx FastLoad, the Final Cartridge, and the Action Replay plug-in ROM cartridges, which all have machine code monitor and disk editor software on board as well. The popular Commodore computer magazines of the era also entered the arena with type-in fast-load utilities, with *Compute!'s Gazette* publishing *TurboDisk* in 1985 and *RUN* publishing *Sizzle* in 1987.

Even though each 1541 has its own on-board disk controller and disk operating system, it is not normally possible for a user to command two 1541 drives to copy a disk (one drive reading and the other writing) as with older dual drives like the 4040 and 8050 that were often found with the PET computer, and which the 1541 is backward-compatible with (it can read 4040 disks but not write to them since its internal operating system is similar enough for reading but not for writing). Unfortunately, however, the routines in the previous disk operating system to enable disk copying were removed for the 1541 as it was intended to be a stand-alone unit. Originally, to copy from drive to drive, software running on the C64 was needed and it would first read from one drive into computer memory, then write out to the other. Only later when first, Fast Hack'em, then other disk backup programs, were released, was true drive-to-drive copying possible for a pair of 1541s. The user could then unplug the C64 from the drives (i.e. from the first drive in the daisy chain) and do something else with the computer as the drives proceeded to copy the entire disk. This is not a recommended practice as disconnecting the serial lead from a powered drive and/or computer can result in destruction of one or both of the port chips in the disk drive.

2.9.4 Media

Each side of 170 kB is split into 683 sectors on 35 tracks, each of the sectors holding 256 bytes; the file system made each sector individually rewritable.

However, one track is reserved by DOS for directory and file allocation information (the BAM, block availability map). And since for normal files, two bytes of each physical sector are used by DOS as a pointer to the next physical track and sector of the file, only 254 out of the 256 bytes of a block are used for file contents.

If the disk side was not otherwise prepared with a custom format, (e.g. for data disks), 664 blocks would be free after formatting, giving $664 \times 254 = 168,656$ bytes (or almost 165 kB) for user data.

By using custom formatting and load/save routines (sometimes included in third-party DOSes, see below), all of the mechanically possible 40 tracks can be used. The reason Commodore decided not to use the upper five tracks by default (or at least more than 35) was the bad quality of some of the drive mechanisms which did not always work reliably at the highest tracks. So by reducing the number of tracks used and thus capacity, it was possible to further reduce cost - in contrast to single-density drives used e.g. in IBM PC computers of the day which save 180 kB on one side (by using a 40-track format). The 1983 Apple FileWare mini-floppy drives use double-sided media, higher track pitch, and variable motor speed to achieve a storage capacity of 871 kB, or 435 kB per side.

The 1541 does not have an index hole sensor, making it straightforward to use the reverse side of a disk by flipping it. A disk can be converted to a "flippy disk" by simply cutting/punching a notch on the left-hand side, causing the drive to recognize both sides as writable. This would effectively double the storage capacity. The notch can be made with scissors, a knife, hole punch, or a disk notcher tool that is specifically designed for this task. Most soft-sectored and all hard-sectored drives would have also required an extra cut-out for the index hole — a harder modification.

Tracks 36-42 are non-standard. The bitrate is after GCR encoding, so actual data is a factor 5/4 less.[6]

The 1541 disk typically has 35 tracks. Track 18 is reserved; the remaining tracks are available for data storage. The header is on 18/0 (track 18, sector 0) along with the BAM (block availability map), and the directory starts on 18/1 (track 18, sector 1). The file interleave is 10 blocks, while the directory interleave is 3 blocks.

Header contents: The header is similar to other Commodore disk headers, the structural differences being the BAM offset ($04) and size, and the label+ID+type offset ($90).

$00–01 T/S reference to first directory sector (18/1) 02 DOS version ('A') 04-8F BAM entries (4 bytes per track: Free Sector Count + 24 bits for sectors) 90-9F Disk Label, $A0 padded A2-A3 Disk ID A5-A6 DOS type ('2A')

2.9.5 Uses

Early copy protection schemes deliberately introduced read errors on the disk, the software refusing to load unless the correct error message is returned. The general idea was that simple disk-copy programs are incapable of copying the errors. When one of these errors is encountered, the disk drive (as do many floppy disk drives) will attempt one or more reread attempts after first resetting the head to track zero. Few of these schemes had much deterrent effect, as various software companies soon released "nibbler" utilities that enabled protected disks to be copied and, in some cases, the protection removed.

Commodore copy protection sometimes depends on specific hardware configurations. *Gunship*, for example, does not load if a second disk drive or printer is connected to the computer.[7]

2.9.6 Notes

[1] Anderson, John J. (March 1984). "Commodore". *Creative Computing*. p. 56. Retrieved 6 February 2015.

[2] "RUN Magazine issue 28".

[3] Levitan, Arlan (December 1988). "Levitations". *Compute!*. p. 104. Retrieved 10 November 2013.

[4] "Physical Exam". *Info*. May–June 1986. p. 57. Retrieved 6 October 2013.

[5] http://www.binarydinosaurs.co.uk/Museum/Commodore/c64/c64notes.php

[6] "Power20 Documentation - File Formats, Appendix E: Emulator File Formats". infinite-loop.at.

[7] Bobo, Ervin (February 1988). "Project: Stealth Fighter". *Compute!*. p. 51. Retrieved 10 November 2013.

2.9.7 References

- CBM (1982). *VIC-1541 Single Drive Floppy Disk User's Manual*. 2nd ed. Commodore Business Machines, Inc. P/N 1540031-02.

- Neufeld, Gerald G. (1985). *1541 User's Guide. The Complete Guide to Commodore's 1541 Disk Drive.* Second Printing, June 1985. 413 pp. Copyright © 1984 by DATAMOST, Inc. (Brady). ISBN 0-89303-738-9.

- Immers, Richard; Neufeld, Gerald G. (1984). *Inside Commodore DOS. The Complete Guide to the 1541 Disk Operating System.* DATAMOST, Inc & Reston Publishing Company, Inc. (Prentice-Hall). ISBN 0-8359-3091-2.

- Englisch, Lothar; Szczepanowski, Norbert (1984). *The Anatomy of the 1541 Disk Drive.* Grand Rapids, MI: Abacus Software (translated from the original 1983 German edition, Düsseldorf: Data Becker GmbH). ISBN 0-916439-01-1.

2.9.8 External links

- C64 Preservation Project: internal drive mechanics and copy protection

- Undocumented 1541 drive functions from the Project 64 website

- RUN Magazine Issue 64

- devili.iki.fi: Beyond the 1541, Mass Storage For The 64 And 128, COMPUTE!'s Gazette, issue 32, February 1986 (market overview)

- This article is based on material taken from 1541 at the Free On-line Dictionary of Computing prior to 1 November 2008 and incorporated under the "relicensing" terms of the GFDL, version 1.3 or later.

2.10 Commodore 1570

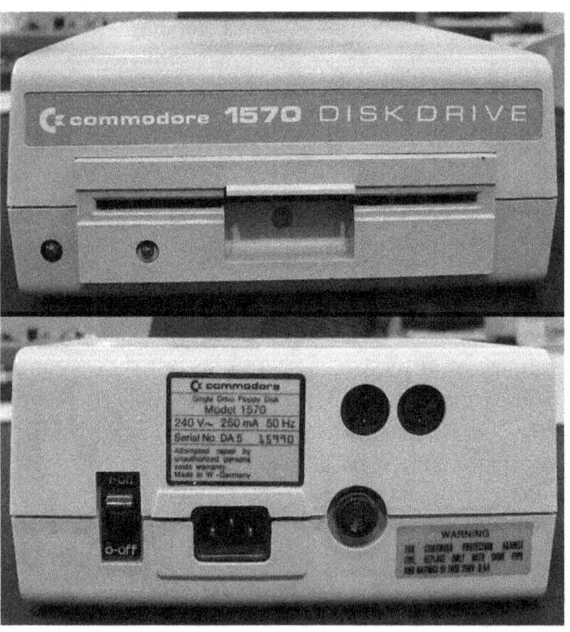

The **Commodore 1570** is a 5¼" floppy disk drive for the Commodore 128 home/personal computer. It is a single-sided, 170-KB version of the Commodore 1571, released as a stopgap measure when Commodore International was unable to provide large enough quantities of 1571s due to a shortage of double-sided drive mechanisms (supplied from an outside manufacturer). Like the 1571, it can read and write both GCR and MFM disk formats. The 1570 utilizes a 1571 logic board in a cream-colored original-1541-like case with a drive mechanism similar to the 1541's except that it was equipped with track-zero detection. Like the 1571, its built-in DOS provides a data burst mode for transferring data to the C128 computer at a faster speed than a 1541 can. Its ROM also contains some DOS bug fixes that didn't appear in the 1571 until much later. The 1570 can read and write all single-sided CP/M-format disks that the 1571 can access.

Although the 1570 is compatible with the Commodore 64, the C64 isn't capable of taking advantage of the drive's higher-speed operation, and when used with the C64 it's little more than a pricier 1541. Also, many early buyers of the C128 chose to temporarily make do with a 1541 drive, perhaps owned as part of a previous C64 setup, until the 1571 became more widely available.

2.11   Commodore 1571

*The **Commodore 1571** disk drive*

The **Commodore 1571** is Commodore's high-end 5¼" floppy disk drive. With its double-sided drive mechanism, it has the ability to use double-sided, double-density (DS/DD) floppy disks natively. This is in contrast to its predecessors, the 1541 and 1570, which can fully read and write such disks only if the user manually flipped them over to access the second side. Because flipping the disk also reverses the direction of rotation, the two methods are not interchangeable; disks which had their back side created in a 1541 by flipping them over would have to be flipped in the 1571 too, and the back side of disks written in a 1571 using the native support for two-sided operation could not be read in a 1541.

2.11.1   Release & features

The 1571 was released to match the Commodore 128, both design-wise and feature-wise. It was announced in the summer of 1985, at the same time as the C128, and became available in quantity later that year. The later C128*D* had a 1571-compatible drive integrated in the system unit. A

double-sided disk on the 1571 would have a capacity of 340 kB (70 tracks, 1,360 disk blocks of 256 bytes each); as 8 kB are reserved for system use (directory and block availability information) and, under CBM DOS, 2 bytes of each block serve as pointers to the next logical block, 254 x 1,328 = 337,312 B or about 329.4 kB were available for user data. (However, with a program organizing disk storage on its own, all space could be used, e.g. for data disks.)

The 1571 features a "burst mode" when used in conjunction with the C128 (although not when used with the Commodore 64 (without modifying hardware) or VIC-20). This mode replaced the slow bit-banging serial routines of the 1541 with a true serial shift register implemented in hardware, thus dramatically increasing the drive speed. Although this originally had been planned when Commodore first switched from the parallel IEEE-488 interface to a custom serial interface, hardware bugs in the VIC-20's 6522 VIA shift register prevented it from working properly.[1]

For compatibility with copy-protected software, the 1571 could closely emulate the 1541. This mode was the default when the drive was used in conjunction with a C64; while always being able to read and write the 1541's GCR format of 170 kB DD single-sided, in this mode it also would format disks single-sided and transfer data at 1541 speed. An undocumented command allowed the drive to format and use the second side of a disk, but only in single-sided mode.

The 1571 was noticeably quieter than its predecessor and tended to run cooler as well, even though, like the 1541, it had an internal power supply (later Commodore drives, like the 1541-II and the 3½" 1581, came with external power supplies). The 1541-II/1581 power supply makes mention of a 1571-II, hinting that Commodore may have intended to release a version of the 1571 with an external power supply. However, no 1571-IIs are known to exist. The embedded OS in the 1571 was CBM DOS V3.0 1571, an improvement over the 1541's V2.6.

Early 1571s had a bug in the ROM-based disk operating system that caused relative files to corrupt if they occupied both sides of the disk. A version 2 ROM was released, but though it cured the initial bug, it introduced some minor quirks of its own - particularly with the 1541 emulation. Curiously, it was also identified as V3.0.

As with the 1541, Commodore initially could not meet demand for the 1571, and that lack of availability and the drive's relatively high price (about US$300) presented an opportunity for cloners. Two 1571 clones appeared, one from Oceanic and one from Blue Chip, but legal action from Commodore quickly drove them from the market.

Commodore announced a dual-drive version of the 1571, to be called the **1572**, but quickly canceled it, reportedly due to technical difficulties with the 1572 DOS. It would have had four times as much RAM as the 1571, and twice as much ROM. The 1572 would have allowed for fast disk backups of non-copy-protected media, much like the old 4040, 8050, and 8250 dual drives.

The 1571 built into the European plastic-case C128 D computer is electronically identical to the stand-alone version, but 1571 version integrated into the later metal-case C128 D (often called C128 DCR, for D Cost-Reduced) differs a lot from the stand-alone 1571. It includes a newer DOS, version 3.1, replaces the MOS Technology CIA interface chip, of which only a few features were used by the 1571 DOS, with a very much simplified chip called 5710, and has some compatibility issues with the stand-alone drive. Because this internal 1571 does not have an unused 8-bit input/output port on any chip, unlike most other Commodore drives, it is not possible to install a parallel cable in this drive, such as that used by SpeedDOS, Dolphin DOS and some other fast third-party Commodore DOS replacements.

2.11.2 Disk format

Unlike the 1541, which was limited to GCR formatting, the 1571 could do both GCR and MFM disk formats. A C128 in CP/M mode equipped with a 1571 was capable of reading and writing floppy disks formatted for many CP/M computers; specifically, the following formats:

- IBM PC CP/M-86
- Osborne 1
- Epson QX10
- Kaypro II, IV
- CBM CP/M FORMAT SS
- CBM CP/M FORMAT DS

Other MFM formats were possible if their characteristics were added to the CP/M C128-specific source code (available from Commodore) and the CP/M operating system were re-assembled. However, booting CP/M was only supported from disks in the standard Commodore GCR format; the MFM formats could only be used once the system was running.

Depending on format, CP/M disks would format to 360 kB, with a mechanical maximum capacity of a 400 kB format (as with DD 5.25" drives generally).

With additional software, it was possible to read and write to MS-DOS-formatted floppies as well. Numerous commercial and public-domain programs for this purpose became available, the best-known being SOGWAP's "Big Blue Reader". Although the C128 could not run any DOS-based software, this capability allowed data files to be exchanged with PC users. Reading Atari 8-bit 130 kB or 180 kB disks was possible as well with special software, but the standard Atari 8-bit 90 kB format, which used FM rather than MFM encoding, could not be handled by the 1571 hardware without modifying the drive circuitry as the control line that determines if FM or MFM encoding is used by the disc controller chip was permanently wired to ground (MFM mode) rather than being under software control.

In the 1541 format, while 40 tracks are possible for a 5.25" DD drive like the 154x/157x, only 35 tracks are used. Commodore chose not to use the upper five tracks by default (or at least to use more than 35) due to the bad quality of some of the drive mechanisms, which did not always work reliably on those tracks. By reducing the number of tracks used (and thus the capacity), Commodore could further reduce cost - in contrast to the double-density drives used e.g. in IBM PCs of the day which saved 180 kB on one side (by using a 40-track format).

For compatibility and ease of implementation, the 1571's double-sided format of one logical disk side with 70 tracks was created by putting together the lower 35 physical tracks on each of the physical sides of the disk rather than using two times 40 tracks, even though there were no more quality problems with the mechanisms of the 1571 drives.

2.11.3 External links

- C64 Preservation Project Discusses internal drive mechanics and copy protection
- RUN Magazine Issue 64
- A photo of the 1572 dual drive, with a 1571 single drive shown for comparison
- The 1572 drive as shown on the Commodore Kuriositäten page (German)
- Information page about the Commodore 1572 (German)
- Secret Weapons of Commodore: The Disk Drives
- Beyond The 1541: Mass Storage for the 64 and 128

2.11.4 References

- Ellinger, Rainer (1986). *1571 Internals.* Grand Rapids, MI: Abacus Software (translated from the original German edition, Düsseldorf: Data Becker GmbH). ISBN 0-916439-44-5.

[1] "Binary Dinosaurs - C64 Notes". 1994-04-07. Retrieved 2013-06-27.

2.12   Commodore 1581

Commodore 1581

The **Commodore 1581** is a 3½-inch double-sided double-density floppy disk drive that was first made by Commodore Business Machines (CBM) in 1987, primarily for its C64 and C128 home/personal computers. The drive stores 800 kilobytes using an MFM encoding[1] but format different from MS-DOS (720 kB), Amiga (880 kB), and Mac Plus (800 kB) formats. With special software it's possible to read C1581 disks on an x86 PC system, and likewise, read MS-DOS and other formats of disks in the C1581 (using Big Blue Reader), provided that the PC or other floppy handles the "720 kB" size format.[1] This capability was most frequently used to read MS-DOS disks. The drive was released in the summer of 1987 and quickly became popular with bulletin board system (BBS) operators and other users.

Like the 1541 and 1571, the 1581 has an onboard MOS Technology 6502 CPU with its own ROM and RAM, and uses a serial version of the IEEE-488 interface. Inexplicably, the drive's ROM contains commands for parallel use, although no parallel interface was available. Unlike the 1571, which is nearly 100% backward-compatible with the 1541, the 1581 has limited compatibility with Commodore's earlier drives. Although it responds to the same DOS commands, most disk utilities written prior to 1987—most notably fast loaders—are so 1541-specific that they do not work with the 1581.

The version of Commodore DOS built into the 1581 added support for partitions, which could also function as fixed-allocation subdirectories. PC-style subdirectories were rejected as being too difficult to work with in terms of block availability maps, then still much in vogue, and which for some time had been the traditional way of inquiring into block availability. When used together with the C128, it implements faster burst mode access than the Commodore 1571 5¼" drive. When using the 1581 together with the C64, however, it is almost as slow as the 1541 drive, due to limitations of the C64's ROM code. The 1581 provides a total of 3160 blocks free when formatted (a block being equal to 256 bytes). The number of permitted directory entries was also increased, to 296 entries. With a storage capacity of 800 kB, the 1581 is the highest-capacity serial-bus drive that was ever made by Commodore (the 1-MB SFD-1001 uses the parallel IEEE-488), and the only 3½" one. However, starting in 1991, Creative Micro Designs (CMD) made the FD-2000 high density (1.6 MB) and FD-4000 extended density (3.2 MB) 3½" drives, both of which offered not only a 1581-emulation mode but also 1541- and 1571-compatibility modes.

Like the 1541 and 1571, a nearly identical job queue is available to the user in zero page (except for job 0), providing for exceptional degrees of compatibility.

Unlike the cases of the 1541 and 1571, the low-level disk format used by the 1581 is similar enough to the MS-DOS format as the 1581 is built around a WD1770 FM/MFM floppy controller chip. PC floppy controllers directly connected via the ISA-bus or onboard, but not standalone USB floppy drives, are able to deal with the 1581 format without need for any special tricks. Thus, utilities to format, read, and write 1581-format disks in standard PC floppy drives under Linux and Microsoft Windows exist. This controller chip, however, was the seat of some early problems with 1581 drives when the first production runs were recalled due to a high failure rate; the problem was quickly corrected. Later versions of the 1581 drive have a smaller, more streamlined-looking external power supply provided with them.

2.12.1   Specifications

This article is based on material taken from the Free On-line Dictionary of Computing prior to 1 November 2008 and incorporated under the "relicensing" terms of the GFDL, version 1.3 or later.

1581 Image Layout

The 1581 disk has 80 logical tracks, each with 40 logical sectors (the actual physical layout of the diskette is abstracted and managed by a hardware translation layer). The directory starts on 40/3 (track 40, sector 3). The disk header is on 40/0, and the BAM (block availability map) resides on 40/1 and 40/2.

Header Contents

$00–01 T/S reference to first directory sector (40/3) 02 DOS version ('D') 04-13 Disk Label, $A0 padded 16-17 Disk ID 19-1A DOS type ('3D')

BAM Contents, 40/1

$00–01 T/S to next BAM sector (40/2) 02 DOS version ('D') 04-05 Disk ID 06 I/O byte 07 Autoboot flag 10-FF BAM entries for Tracks 1-40

BAM Contents, 40/2

$00–01 00/FF 02 DOS version ('D') 04-05 Disk ID 06 I/O byte 07 Autoboot flag 10-FF BAM entries for Tracks 41-80

2.12.2   See also

- Commodore 128

2.12.3   References

[1] tds.net - 1581 DISK DRIVE DIAGNOSTICS, latest updates and corrections 7-13-05

2.12.4   External links

- d81.de: Permanent home of 1581-Copy, A MS-Windows based Tool uses any standard x86-PC 3.5" drive to WRITE & READ 1581 disk images (d81).

- optusnet.com.au: 1581 Games, Commodore 1581 Games, D81 , CMD FD2000 & FD4000 Games, Tools & Games specifically for the 1581 disk drive.

- optusnet.com.au: SEGA SF-7000 with PC 3.5" Floppy Drive, Copy disk to PC and vise versa, How to use a PC 3.5" floppy drive in the 1581 device

- vice-emu: Commodore compatible Disk Drives, drive info

- tut.fi: DCN-2692 floppy controller board, C1581 clone (complete)

2.13   Commodore 64 disk / tape emulation

Commodore 64 disk/tape emulation and data transfer comprises hardware and software for Commodore 64 disk & tape emulation and for data transfer between either Commodore 64 (C64), Commodore (1541) disk drive or Commodore (1530 Datasette) tape deck and newer computers.

There is a large variety of adapters for C64 disk/tape emulation and data transfer, and an even larger variety of compatible software. Many of the adapters are interfacing the original *serial bus* disk drive plug or the *C2N* tape interface. Other connects to either the *user port* or the cartridge *expansion port* using either standardized RS232 interfaces or proprietary adapters. In combination with software (or firmware), the adapters can either fully support the original communication protocols, provide partial support or apply proprietary communication protocols. Different solutions allows for letting the C64 access programs stored on another computer or the Internet, and for accessing the C64 disk station and tape deck from other computers.

2.13.1   History

Some of the oldest adapters are the *C64 user port* to RS232 converters. Those were standardized and originally designed to connect printers and other 3rd party hardware, including modems. Later, those adapters have also been adopted for disk drive emulation and even Internet connections. However, the most widespread adapters were probably the different disk drive and printer plug *C64 serial bus* to Parallel port adapter that evolved for transferring data between disk drives and parallel port supplied computers. Because of hard timing requirements on the C64 side, those are unfortunately not applicable to laptops or multitasking operation systems. There also exist a more limited number of adapters for the *C64 tape interface*. While the data transfer over the *user port* is usually limited to 2.4 kbit/s, the *C64 expansion port* cartridge interface supports transfer rates of one to two magnitudes higher through proprietary protocols. There exist *C64 expansion port* adapters that support both *hard disks*, *memory cards*, *USB-disks* and *Ethernet* connections.

The software is typically open source, and so is most of the hardware designs. You can therefore build most of the hardware yourself, though they are usually also available from online shops.

Software for C64 disk & tape emulation

This section comprises software for emulating the 1541 disk drive or the Commodore 1530 Datasette tape deck on external computers, making them available to a physical Commodore 64.

.* no software required with C2N232, but with X1541 and

PC64

.** a simple Basic version of the software on the C64 side is available for typing in to the C64 before the first use.

Software for PC to disk & tape transfer

This section comprises software for transferring files and images between the 1541 disk drive or the Commodore 1530 Datasette tape deck and an external computer different from the Commodore 64.

.*This software requires that OpenCBM is available.

Hardware

The following table addresses hardware for connecting the Commodore 64, the 1541 disk drive or the Commodore 1530 Datasette tape deck to external computers, data storage (such as disks and memory cards) or the Internet.

2.13.2   Software by C64 compatibility

Full emulation of the Commodore 1541 disk drive or Commodore 1530 datasette is required e.g. to support fast loaders. Software that supports the basic transfer protocols, such as load and save, will not support fast loaders.

Software exists that replaces the basic transfer protocols with proprietary alternatives. These protocols require special software on both the host side and the Commodore 64 side.

Some software supports transfer between a disk or tape drive and a computer other than a Commodore 64.

Floppy disks

Full emulation

- The *1541EMU* emulates the internal hardware of the Commodore 1541 disk drive on a host computer and supports the 1541EMU cable only. The 1541EMU software was designed in 2001-2002 by Ville Muikkula et al.[3] The realtime requirements for emulating the 1541 disk drive are exceptionally hard. PCs with multitasking operating systems are therefore not supported. Even though the timing can be adjusted in software, the communication is not likely to work on PCs with newer processors than Pentium.[ref required]

- The *1541 Ultimate* firmware completely emulates the Commodore 1541 disk drive on the 1541U-I cartridge, using a disk connector output on the cartridge.

Full emulation is also planned for the 1541U-II cartridge. It allows access to the disk content on memory cards (I) and USB-disks (II). The selected disk can be downloaded either through an accurate 1541 on-board emulation (II). A proprietary protocol is also available for faster but not fully compatible transfer (both I and II).[4][5]

- The *Chameleon* firmware completely emulates the Commodore 1541 disk drive on the Turbo Chameleon 64 cartridge, using a disk connector output on the cartridge. The emulator provides access to a MMC or SD card on the cartridge.

Partial emulation

- The *64HDD* is a serial bus and disk drive emulator, developed in 1999-2010 by Nicholas Coplin.[6]

- The *MMC2IEC* firmware provides emulation of the Commodore 1541 disk drive on the MMC2IEC adapter, providing access to a MMC memory card.

- The *sd2iec* firmware provides emulation of the Commodore 1541 disk drive on the SD2IEC adapter, providing access to a SD memory card. It also supports many fast loaders.

- The *uIEC* firmware provides emulation of the Commodore 1541 disk drive on the uIEC adapter, providing access to either an IDE hard disk, a CF memory card or a SD memory card, depending on the version.

- The *VC1541* is a Commodore 1541 serial protocol emulator developed in 1997-1998 by Torsten Paul.[7]

Proprietary options

- The *Over 5* is a software package for transferring files between a C64 and an host machine (which can be an Amiga, PC or Unix box). It works in two different ways, either using the C64 as a server to the host for accessing floppy disks or using the host as a hard disk server for the C64. Over5 was developed in ????–2002 by Daniel Kahlin. It was later ported to Win32 by Martin Sikström and to Unix by Andreas Anderson.[8]

- The *Serial Slave* allows you to use a PC as a virtual disk drive for your C64 or C128. It was developed in 2001-2002 by Per Olofsson and friends.[9]

- The *V-1541* is a program that replaces the standard LOAD and SAVE operations on a Commodore 64 computer. The V-1541 program allows your Commodore 64 computer to access files and other content on the Internet at CommodoreServer.com. CommodoreServer.com is a Virtual Disk Drive to which you can upload D64 disk images from any Internet computer and later download the disk from the Commodore 64.[10]

Disk transfer

- The *C64S* (C64 Software Emulator) is a Commodore 64 emulator that supports transfer to/from a 1541 disk drive. It was developed in 1994-1997 by Miha Peternel.

- The *DISK64* is a disk transfer tool developed in 1993-1994 by Alfred Schwall.

- The *MNIB* was developed in 2000-2004 by Markus Brenner. It has been succeeded by NIBTOOLS.

- The *NIBTOOLS* is a disk transfer program designed for copying original disks and converting them into G64 and D64 disk image formats on a PC. NIBTOOLS requires OpenCBM. NIBTOOLS is based on MNIB and was developed 2005-2010 by Pete Rittwage, Markus Brenner and the C64 Preservation Project team.[11]

- The *OpenCBM* allows for access to a VIC 1540, 1541, 1570, 1571, or even 1581 floppy drive from the PC on Windows NT, 2000 and XP. With OpenCBM you can copy D64 or D71 images from a real drive to the PC, or from the PC to a real drive with the help of *d64copy*. Furthermore, you can copy single files in both directions. Some more tools (for example, cbmctrl) are given, too. OpenCBM started out as *cbm4linux*, a Linux-only solution written in 1999-2003 by Michael Klein.[12] Spiro Trikaliotis ported it over to the Windows platform in 2005 under the name *cbm4win*. With version 0.4.0, both versions were joined back into one source with the common name *OpenCBM*. For Windows, due to limitations of the drivers of the cards, it is unlikely that PCI or CardBus cards will work. However, ISA cards will work. For Linux, chances are high that all cards will work.[13]

- The *Personal C64* is a Commodore 64 emulator that supports transfer to/from 1541 disk drive. It was developed in 1994-1997 by Wolfgang Lorenz.[14]

- The *Star Commander* copies files and disks between a PC and the Commodore 1541/1570/1571/1581 drive, optionally using fast loader. Star Commander was developed in 1994-2010 by Joe Forster/STA.[15] Commodore disk drives expect a tighter synchronization than the Commander can keep under a multi-tasking

PC operating system. The best results is therefore obtained by running the Commander under plain DOS.[16]

- The *Trans64* is a program to transfer files between the PC and a C64 floppy drive. Trans64 was developed in 1994-1997 by Bernhard Schwall.

- The *X1541* was developed in 1992 by Leopoldo Ghielmetti.

Cassette tapes

Full emulation

- The *C2N* fully emulates the Commodore 1530 via the C2N232 device. C2N was developed in ???? by Marko Mäkelä.[17]

- The *C2NLOAD* first provides a turbo tape over the standard Commodore 1530 datasette Load routine, and then automatically loads the requested program file at 38.400 bit/s. C2NLOAD was developed in 2001-2006 by Marko Mäkelä.[18]

Proprietary options

- The *CBMLINK* is a data transfer system between Commodore 8-bit computers and other systems (Amiga, IBM PC compatible, Apple, Unix workstations). Supported by the VICE emulator. CMBLINK was developed in 2001-2003 by Marko Mäkelä, based on PRLINK.[19]

- The *Linux Server 64* was developed by Roger Lawhorn. It supports the same commands as CMBLINK, but adds a series of commands for printing and scanning using hardware connected to a PC.[20]

- The *Prlink* is a software for data transfer between Commodore 8-bit computers and an Amiga (AmigaDOS) or a PC clone (Linux, MS-DOS). Prlink was developed in 1994-1996 by Marko Mäkelä and Olaf Seibert. It was succeeded by CBMLINK developed in 2001-2003 by Marko Mäkelä.[19]

Tape transfer

- The *mtap & ptap* are MS-DOS tools for creating real tape files (.TAP files) from original C64, VIC-20 and C16 tapes using the Commodore Datasette, and for playing back .TAP files to real tapes for use with an actual Commodore 64 machine. mtap & ptap were developed in 1998-2002 by Markus Brenner.[21]

2.13.3   Hardware by C64 compatibility

Tape connector adapters

- The *C2N232* adapter is a RS-232 interface that can be plugged to the cassette port of an 8-bit Commodore computer and supports emulation of the tape deck. The C2N232 hardware was designed in 2001-2003 by Marko Mäkelä. It is freely available as open source, and a few hundred were built and sold.[17]

- The *C64S* tape adapter lets you connect your tape deck to a PC parallel port.[22]

- The *Cassadapt* tape adapter allows to convert tape programs (T64 and PRG) from a PC to either the Commodore 64 or a C2N tape deck.[23]

Disk connector adapters

- The *1541-III* is a PIC microcontroller controlling a MMC/SD card with .D64 files. It does however NOT support fastloaders.[24]

- The *1541EMU* cable hardware supports full emulation of the Commodore 1541 disk drive. The cable exists in type0 and type1 flavors with full and slightly reduced compatibility, respectively.[3] The 1541EMU hardware was designed in 2001-2002 by Ville Muikkula *et al.* for use with the 1541EMU software. The 1541EMU cable is available through both building instructions and shops.[25]

- The *1541U-I* (and *1541U-II*) emulates a 1541 disk drive for Commodore computers. It uses an SD-card or MMC-card to store virtual floppy disks. Disk content can be browsed through software on the cartridge and navigated through buttons on the device. The selected disk can be downloaded either through the fully compatible 1541 interface or through a faster but less compatible proprietary interface.[4]

- The *X1541* cables allow (full emulation of? /) copying to and from the Commodore 1541 disk drive. The realtime requirements for emulating the 1541 disk drive are exceptionally hard, and a variety of cable flavors have been constructed to improve compatibility with multi-tasking systems and faster PCs than the Pentium to some degree.[26] For compatibility, check the documentation of each application, or confer the reference. The X1541 cable is available through both building instructions and shops.[25][27] The original X1541 cable was designed in 1992 by Leopoldo Ghielmetti for use with the X1541 software.

- The *XP1541* cables are variants of the X1541 cables flavors, adding parallel support for faster transfer between PC and the C64 disk drive. The XP1541 cable was designed in 1997 by Joe Forster/STA.[28]

- The *XU1541* adapter (beta) attaches a 1541 etc. disk drives to a PC using the USB connection, opening for easy transfer of disk images from and to the disk drive. The XU1541 is currently only recommended for people who are willing to cope with glitches and will perhaps even do some testing and bug hunting.[29]

- The *ZoomFloppy* connects your Commodore 1541/1571/1581 drives to a Windows, Mac, or Linux computer. This allows you to read and write files or entire disk images from the original media.[30] The ZoomFloppy uses the XUM1541 firmware by Nate Lawson, based on XU1541.[31] The PCB was manufactured by Jim Brain and can also be bought online.[32]

RS232 user port adapters

The User Port RS232 adapters provides a low-speed serial port for Commodore 8-bit computers, originally for connecting printers etc. They can operate at speeds of up to 2.4 kbit/s.

- The *Comet64 Internet Modem* is a Serial-to-Ethernet (S2E) device. This modem connects to the user port and provides an Internet connection. Also available with RS232 output.[33]

- The *EZ-232 RS232* Serial Interface provides a low-speed serial port for Commodore 8-bit computers. It can operate at speeds of up to 2400bit/s, when configured as a standard interface, and at speeds of up to 9600bit/s, when configured as a UP9600 interface. The EZ-232 RS232 Serial Interface was designed by Jim Brain.[34]

- The *Handic V24* is a RS-232 converter for Commodore machines.

- The *VIC-1011A RS232C* is a RS-232 converter for Commodore machines.[35]

Proprietary user port adapters

- The *PC64* cable was designed in 1994 by Wolfgang Lorenz.[36]

- The *Power-Loader* cable is a companion for the X1541 cables flavors, adding parallel support for faster transfer between PC and the C64. The Power-Loader (Pwr/XE) cable was developed by Nicholas Coplin, for use with 64HDDsoftware.[37]

- The *Prlink* is a cable designed for the Prlink software by Marko Mäkelä and Olaf Seibert.

RS232 expansion port adapters Expansion port cartridges provides a high-speed connection to an external computer and/or the Internet. The output interface is a RS232 interface for connecting to an external computer. Internet connection can be obtained through the external computer or via a series to Internet adapter.[38] Some adapters also have separate Ethernet interface for connecting to Internet. The most common cartridges for external connection is listed below.[39]

- The *RR Net* is an add on to the *Retro-Replay* cartridge that allows for broadband Internet access.

- The *Silver Surfer* add on to the *Retro-Replay* cartridge provides RS232 capabilities. It provides a transfer rate probably similar to that of *Turbo232*, limited to 57.8 kbit/s for most software.[40] Silver Surfer is available from an online store.[41]

- The *SwiftLink-232* is a RS-232 serial port cartridge for the C64/128. It provides up to 38.4 kbit/s transfer rate.[42] Swiftlink was manufactured by CMD, who has stopped producing Commodore gear. There exist several clones such as the *Datapump cartridge* or the *Pitchlink cartridge*.[43] The *Link-232* is another clone of Swiftlink.

- The *Turbo232* cartridge is a high-speed RS-232c modem interface for Commodore 64 or 128 computer. It provides up to 230 kbit/s transfer rate, though limited to 57.8 kbit/s for most software.[44] Turbo232 was manufactured by CMD, who has stopped producing Commodore gear.

Proprietary expansion port adapters

- The *1541Ultimate-II* emulates a 1541 disk drive for Commodore computers on a cartridge, using MicroSD or USB disks to store virtual floppy disks. The disk can be downloaded through fast, but not fully compatible proprietary disk emulation. Disk connector for fully compatibility is integrated, but not yet supported in software (as for the 1541U-I). Disk content can be browsed through software on the cartridge and navigated through buttons on the device.[4]

- The *Comet64 Internet Modem* Serial-to-Ethernet device provides an Ethernet connection. See the Comet64 Internet Modem entry under the *User Port RS232* section.

2.13.4 References

[1] *Cable*: Wires and plugs only, *Connector*: Wires, plugs and simple components, *Adapter / Cartridge*: PCB with standard and or programmable components

[2] Disk connector interfaces (e.g. x1541/xp1541 cables) have hard real-time requirements, and are in general not compatible with Pentium and newer PCs or multitasking operation systems.

[3] "Home of 1541EMU". Kotinet.com. Retrieved 2013-10-11.

[4] "Home of 1541 Ultimate". 1541ultimate.net. Retrieved 2013-10-11.

[5] "1541 Ultimatec on the 64-Wiki". C64-wiki.com. 2010-11-30. Retrieved 2013-10-11.

[6] "Home of the 64HDD project". 64hdd.com. Retrieved 2013-10-11.

[7] "Home of VC1541". Vc1541.sourceforge.net. 2003-07-29. Retrieved 2013-10-11.

[8] "Home of Over5". Kahlin.net. Retrieved 2013-10-11.

[9] Home of Serial Slave

[10] "Official blog of Commodore Server". Commodore-server.com. Retrieved 2013-10-11.

[11] "NIBTOOLS at the C64 Preservation Project". C64preservation.com. Retrieved 2013-10-11.

[12] "Home of cbm4linux". Lb.shuttle.de. Retrieved 2013-10-11.

[13] "OpenCBM on Spiro's home on the web". Trikaliotis.net. Retrieved 2013-10-11.

[14] "Home of Personal C64". Zimmers.net. 1997-06-05. Retrieved 2013-10-11.

[15] "Home of Star Commander". Sta.c64.org. 2010-01-11. Retrieved 2013-10-11.

[16] Documentation of StarCommander, section 3

[17] "Marko Mäkelä's old computers: Commodore C2N Datasette Codec c2n". Ktverkko.fi. Retrieved 2013-10-11.

[18] "Marko Mäkelä's old computers: Commodore C2N datassette emulator with RS-232 interface". Ktverkko.fi. Retrieved 2013-10-11.

[19] "Marko Mäkelä's old computers: cbmlink". Zimmers.net. Retrieved 2013-10-11.

[20] "Home of Linux Server 64". Rll.home.insightbb.com. Retrieved 2013-10-11.

[21] "Minstrel's Commodore page: mtap & ptap". Markus.brenner.de. Retrieved 2013-10-11.

[22] "Markus Brenner: The C64S tape adapter". Markus.brenner.de. Retrieved 2013-10-11.

[23] "C8D Cassadapt". Cbm8bit. Retrieved 2013-10-11.

[24] "1541-III". Jderogee.tripod.com. Retrieved 2013-10-11.

[25] "The X1541 Shop". Sta.c64.org. Retrieved 2013-10-11.

[26] "The X-series interfaces". Sta.c64.org. Retrieved 2013-10-11.

[27] "64Copy Central". Ist.uwaterloo.ca. 2012-01-12. Retrieved 2013-10-11.

[28] "The XP1541 interface". Sta.c64.org. Retrieved 2013-10-11.

[29] "Spiro's home on the web: The XU1541". Trikaliotis.net. 2008-01-28. Retrieved 2013-10-11.

[30] "Nate Lawson: The ZoomFloppy manual" (PDF). Retrieved 2013-10-11.

[31] "Nate Lawson: The XUM1541 firmware". Root.org. Retrieved 2013-10-11.

[32] IEEE Connector:. "Go4Retro online store: ZoomFloppy". Store.go4retro.com. Retrieved 2013-10-11.

[33] "The Comet64 Internet Modem". Commodoreserver.com. Retrieved 2013-10-11.

[34] "Everything Commodore The EZ-232 manual". Retrieved 2013-10-11.

[35] "Denial Wiki: VIC-1011A RS232C adapter". Sleepingelephant.com. 2012-01-25. Retrieved 2013-10-11.

[36] "The Joe Forster/STA: PC64 overview". Sta.c64.org. Retrieved 2013-10-11.

[37] "The Joe Forster/STA: Power-Loader". Sta.c64.org. Retrieved 2013-10-11.

[38] "Using a UDS-10 (UDS-100) Device Server". Armory.com. Retrieved 2013-10-11.

[39]

[40] "Silver Surfer". Ar.c64.org. Retrieved 2013-10-11.

[41] Vesalia.de online store: Silver Surfer.

[42] Replay Resources: The SwiftLink manual

[43] "Andreas Andersson: Over5 and Swiftlink clones". E.kth.se. 2002-04-21. Retrieved 2013-10-11.

[44] Replay Resources: The Turbo232 manual

2.13.5 External links

- Data transfer alternatives by *Bo Zimmerman* (updated around 2005): Data Transfers with Commodore Computers

- Disk and tape transfer at *The Starcommander* pages (updated around 2010) Useful DOS software, Useful Windows software, Useful Linux software, etc.

- Relevant tools listed by *World of Fairlight* at The C64 tool list.

- Tape transfer and conversion (updated around 2006) Alternatives to *WAV-PRG and Audiotap*.

- Internet connection: The Internet For Commodore C64/128 Users - a manual from Commodore Homestead.

2.14 Commodore 64 Games System

The **Commodore 64 Games System** (often abbreviated **C64GS**) is the cartridge-based home video game console version of the popular Commodore 64 home computer. It was released in December, 1990 by Commodore into a booming console market dominated by Nintendo and Sega. It was only released in Europe and was a considerable commercial failure.

The C64GS came bundled with a cartridge with four games: *Fiendish Freddy's Big Top O'Fun*, *International Soccer*, *Flimbo's Quest* and *Klax*.

The C64GS was not Commodore's first gaming system based on the C64 hardware. However, unlike the 1982 MAX Machine (a game-oriented computer based on a very cut-down version of the same hardware family), the C64GS is internally very similar to the complete C64, with which it is compatible.

2.14.1 Available software

Support from games companies was limited, as many were unconvinced that the C64GS would be a success in the console market. Ocean Software was the most supportive, offering a wide range of titles, some C64GS cartridge-based only, offering features in games that would have been impossible on cassette-based games, others straight ports of games for the original C64. Domark and System 3 also released a number of titles for the system, and conversions of some Codemasters and MicroProse games also appeared. Denton Designs also released some games, among them Bounces, which was released in 1985.

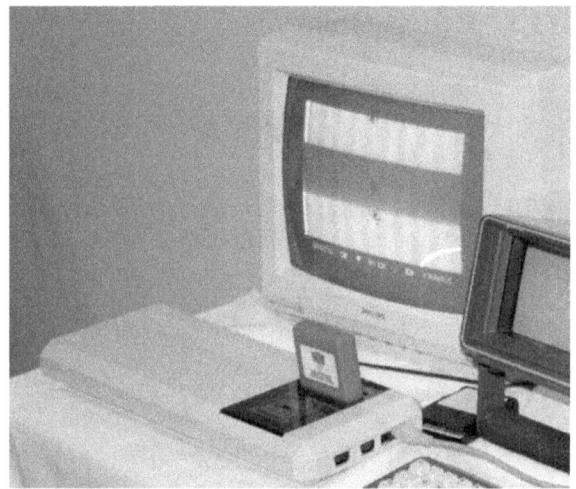

The software bundled with the C64GS, a four-game cartridge containing *Fiendish Freddy's Big Top O'Fun*, *International Soccer*, *Flimbo's Quest* and *Klax*, were likely the most well-known on the system. These games, with the exception of *International Soccer*, were previously ordinary tape-based games, but their structure and control systems (no keyboard needed) made them well-suited to the new console. *International Soccer* was previously released in 1983 on cartridge for the original C64 computer.

Ocean produced a number of games for the C64GS, among them a remake of *Double Dragon* (which seemed to be more linked to the NES version than the original C64 cassette version), *Navy SEALS*, *Robocop 2*, *Robocop 3*, *Chase HQ 2: Special Criminal Investigation*, *Pang*, *Battle Command*, *Toki*, *Shadow of the Beast* and *Lemmings*. They also produced *Batman The Movie* for the console, but this was a direct conversion of the cassette game, evidenced by the screens prompting the player to "press PLAY" that briefly appeared between levels. Some of the earliest Ocean cartridges had a manufacturing flaw, where the connector was placed too far back in the cartridge case. The end result was that the cartridge could not be used with the standard C64 computer. Members of Ocean staff had to manually drill holes in the side of the cartridges to make them fit.

System 3 released *Last Ninja Remix* and *Myth: History in the Making*, although both were also available on cassette. Domark also offered two titles, *Badlands* and *Cyberball*, which were available on cartridge only.

Through publisher The Disc Company a number of Codemasters and MicroProse titles were also reworked and released as compilations for the C64GS. *Fun Play* featured three Codemasters titles: *Fast Food Dizzy*, *Professional Skateboard Simulator* and *Professional Tennis Simulator*. *Power Play* featured three MicroProse titles: *Rick Dangerous*, *Stunt Car Racer* and *MicroProse Soccer*, although *Rick*

Dangerous was produced by Core Design, not MicroProse themselves. *Stunt Car Racer* and *MicroProse Soccer* needed to be heavily modified to enable them to run on the C64GS.

Uncharacteristically, Commodore never produced or published a single title for the C64GS beyond the bundled four-game cartridge. *International Soccer* was the only widely available game for the C64GS but had actually been written for the C64.

2.14.2 Hardware-based problems

The C64GS has been plagued with problems from the outset. Firstly, despite the wealth of software already available on cartridge for C64, the lack of a keyboard means that most cannot be used with the console. This means that people had often bought secondhand C64 software on cartridge only to find that the games are not compatible. The standard C64 version of *Terminator 2: Judgment Day* was designed for the console, but was included on a cartridge that required the user to press a key to access the game, rendering it unplayable.

To partially counter the lack of a keyboard, the basic control system for the C64GS was a joystick supplied by Cheetah called the Annihilator. This joystick, while using the standard Atari 9-pin plug, offers two independent buttons, with the second button located on the base of the joystick. This 9-pin plug was standard of many systems of the era, and the joysticks are fundamentally compatible with the ZX Spectrum's Kempston Interface and the Sega Master System. The Cheetah Annihilator joystick was poorly built, had a short life, and was not widely available, making replacements difficult to come by.

2.14.3 Primary reasons for failure

Prior to the console's release, Commodore had generated a great deal of marketing hype to generate interest in an already crowded market. *Zzap! 64* and *Your Commodore*, Commodore 64 magazines of the era, reported that Commodore had promised "up to 100 titles before December",[2] even though December was two months from the time of writing. In reality 28 games were produced for the console during its shelf life - most of which were compilations of older titles, and a majority of which were from Ocean. Of those 28 titles, only 9 were cartridge-exclusive titles, the remainder being ports of older cassette-based games.

While most of the titles that Ocean announced did appear for the GS (with the notable exception of *Operation Thunderbolt*), a number of promises from other publishers failed to materialize. Although Thalamus, The Sales

Curve, Mirrorsoft and Hewson had expressed an interest, nothing ever materialized from these firms. Similar problems plagued rival company Amstrad when they released their GX4000 console the same year.

There were other reasons attributed to the failure of the C64GS, the major ones being the following:

- *Poor software support:* Most existing software on cartridge did not function well with the C64GS, and enthusiasm from publishers was low. Ocean Software, Codemasters, System 3, MicroProse and Domark developed titles for the system, but probably only because the games were compatible with the original C64, providing the titles with a commercial safety net in case the C64GS failed. And failure to reprogram the games for use with the cut-back system was another blame for the fault.

- *The C64 computer:* The C64GS was essentially a cut-back version of the original Commodore 64, and the games developed for it could also be run on the original computer. The C64 was already at an affordable price, and the C64GS was sold for the same. People preferred the original C64, particularly since the cassette versions of games could often be picked up for a fraction of the cost of the cartridge versions.

- *Obsolete technology:* The C64 was introduced in 1982; by 1990 the technology was way past its prime.

- *An already saturated console market:* The 8-bit C64GS entered the market in 1990 parallel to newer fourth generation consoles such as the Mega Drive and the Super Nintendo. The Nintendo Entertainment System and Sega Master System were already dominating the market with more popular titles, and did so until around 1992.

- *TV hookup*, joystick support and cartridge slots were already found on regular C64 machines. Hence normal C64s were already recognized as "game consoles" despite actually being home computers with integrated keyboards.

Commodore eventually shipped the four-game cartridge and Cheetah Annihilator joysticks in a "Playful Intelligence" bundle with standard Commodore 64C computer. Several years later Commodore's next attempt at a games console, the Amiga CD32, encountered many of the same problems although overall it was a lot more successful than the C64GS.

2.14.4   Technical specifications

The specifications of the C64GS are a subset of those of the regular C64; the main differences being the leaving out of the user port, serial interface, and cassette port. Since the system board is a regular C64C board these ports are actually present, but simply not exposed at the rear.

Internal hardware

- Microprocessor CPU:

  - MOS Technology 8500 (the 6510/8500 being a modified 6502 with an integrated 6-bit I/O port)
  - Clock speed: 0.985 MHz (PAL)

- RAM:

  - 64 KB (65,535 bytes).
  - 0.5 KB Color RAM

- ROM:

  - 20 KB (7 KB KERNAL, 4 KB character generator providing two 2 KB character sets)

The ROM contains two important differences from a standard C64 ROM. The first is that switching on the machine without a cartridge present results in a character-based animation asking the user to insert a cartridge. The second is an additional set of windowing commands, designed to compensate for the lack of a keyboard. However, there is no known software that uses it.

- Video hardware: MOS Technology VIC-II MOS 8569 (PAL)

  - 16 colors
  - Text mode: 40×25; user-defined characters; smooth scrolling
  - Bitmap modes: 320×200, 160×200 (multicolor)
  - 8 hardware sprites, 24×21 pixels

- Sound hardware: MOS Technology 8580 "SID"

  - 3 voices, ADSR programmable.
  - 4 Waveforms: Triangle, Sawtooth, Variable Pulse, Noise
  - Oscillator Synchronization, Ring modulation
  - Programmable Filter: High Pass, Low Pass, Band Pass, Notch Filter

I/O and power supply

- I/O ports:

  - High-quality Y/C (S-Video) (8-pin DIN plug) with chroma/luma out and sound in + out, used with some Commodore video monitors (DIN-to-phono plug converter delivered with monitor).

  - Composite video (one-signal video output to monitor included in aforementioned 8-pin DIN plug, and separate integrated RF modulator antenna output, which also carries sound, to TV on an RCA socket)

  - 2 × screwless DE9M game controller ports (Atari 2600 de facto standard, supporting one digital joystick each)

  - Cartridge slot (44-pin slot for edge connector with 6510 CPU address/data bus lines and control signals, as well as GND and voltage pins;[3] used for program modules)

- Power supply: 5V DC and 9V AC from external "monolithic power brick", attached to computer's 7-pin female DIN-connector

2.14.5 See also

- Commodore 64

- Commodore MAX Machine

2.14.6 References

[1] "Retro Treasures: Commodore 64 GS". retro-treasures.blogspot.se. 2013-06-07. Retrieved 2013-06-16.

[2] "Munchy Box". *Your Commodore* (UK: Alphaville Bublications Ltd.). October 1990.

[3] The Hardware Book

2.14.7 External links

- "The C64 Console!" / "Inside the future: The C64GS" – By Ed Stu, *Zzap 64* magazine, issue 66, October 1990

- The Commodore C64 Games System – Photos and information from Bo Zimmermann's collection

- 8Bit-Homecomputermuseum – Nice pictures of the C64GS

2.15 Commodore 64 joystick adapters

Commodore 64 joystick adapters are hardware peripherals that extend the number of joystick ports on the Commodore 64 computer. The additional joysticks can be used on games with dedicated support for the specific adapter.

A number of different joystick adapters have been constructed for use with the C64. The *Classical Games / Protovision* adapter is by far supported by the largest number of games. While building instructions are available for most of the adapters, a few adapters are also available commercially. The adapters are also emulated in some of the C64 emulators.

2.15.1 Original Adapters

Starbyte adapter (1990)

Supports 30 additional joysticks at the userport. Released by *Starbyte Software* in 1990 for use with their game *Adidas Championship Tie Break*. At the time sold separately under the name *Tie Break Adaptor*.[1][2]

Building instructions (plugs and wires only) have been reverse engineered and made available by Groepaz/Hitmen.[3]

Kingsoft adapter (1992)

Supports 2 additional joysticks at the userport. Introduced with the game *Bug Bomber* by *Kingsoft* in 1992. The game came with instructions to build your own 4-player adapter.[4]

Building instructions (plugs and wires only) have been reverse engineered and made available by Groepaz/Hitmen.[3]

Protovision 4 player interface / Classical Games adapter (1997)

Supports 2 additional joysticks at the userport. Designed by *Chester Kollschen (CKX)* of *Classical Games* and released in 1997 with the game *Bomb Mania*. Kollschen later joined *Protovision* and the adapter is now available as the *Protovision 4 player interface*.[5][6][7]

Building instructions (plugs, wires and an IC), sample code, shop and list of supported games are available at the Protovision website.[7]

Space Balls adapter (1998)

Supports 8 joysticks, all connected to the user port and one joyport. Designed by *Luigi Pantarotto* and released with the game *Space Balls* (1998) for 2-8 players.

Building instructions (plugs and wires only) is provided with the game *Space Balls* (called *Star Balls* in its schematics page).[4]

Ninjas SNES pad adapter (1998)

Supports 8 SNES pads at the joyports. Designed by *Ninja* of *The Dreams*.[8]

Building instructions (plugs and wires only) are available at the *Hitmen 4 player adapter* website.[3] Code interface is demonstrated in the *SNES-Pad-Tooldisk #1*.[9]

Digital Excess & Hitmen 4 player joystick adapter (1999)

Supports 2 additional joysticks at the userport. Designed by *Groepaz* of *Hitmen* and *Thomas Koncina* and *Bjoern Oden-dahl* of *Digital Excess* in 1999 as an easier to build alternative to the *Classical Games* adapter.[10]

Building instructions (plugs and wires only) and sample code is available at the *Hitmen 4 player adapter* website.[3]

Inception - 8 joysticks adapter (2013)

Supports 8 joystics at one joystick port. Designed by Ray and PCH (Petr Chlud). [11]

Available commercially at the c64.cz website.[12]

2.15.2   Compatible Adapters

Singular Crew 4Player Adapter (2008)

Supports 2 additional joystics at the userport. Compatible with the Protovision 4 player interface / the Classical Games Adapter, Digital Excess & Hitmen 4 player joystick adapter and the Kingsoft Adapter. Designed by Singular Crew. [13]

Available commercially at www.zsibvasar.hu website.[14]

2.15.3   Related Adapters

Adapters for a few related computers can also be used with the C64. Those adapters include *PET* (normally only used on the PET/CBM2), *Hummer* (normally only used on the C64DTV) and *OEM* (normally only used on the VIC20).[15]

There are probably no game supporting those adapters on the C64.

2.15.4   Emulation

The C64 emulator VICE supports the *Classical Games / Protovision* and the *Digital Excess & Hitmen* adapters from version 2.3.,[15] and the *Kingsoft* and *Starbyte* adapters from version 2.4.[16]

2.15.5   Supported games

A (near) complete list of games with original or added support for the various adapters. [3] [7] [17] [18] [19]

Adapter: S = Starbyte, C = Protovision / Classical Games, D = Digital Excess & Hitmen, N = Ninjas SNES pad, K = Kingsoft, B = Space Balls, I = Inception

Description: original/clone = new implementation, patch = patch for existing game, hack = unauthorized release based on existing game.

Interface origin: Publisher or official source of joystick adapter support. In order to use hacked versions legally, you should own the original game.

2.15.6   References

[1] "Zzap!Test! Tie Break", Zzap!64, 1990 Nov (67), p 12. zzap.co.uk. Retrieved 6 March 2011.

[2] "Adidas Championship Tie Break", Ocean Software & Starbyte Software, 1990. Stadium 64 Manuals Archive. Retrieved 6 March 2011.

[3] "4 Player Adapter". Hitmen. Retrieved 6 March 2011.

[4] Synnes, Stig: Joystick adapter schematics

[5] "Commodore Free interview with the Protovision team", Commodore Free, 2011, issue 51, p 37. Commodore Free. Retrieved 8 August 2011.

[6] "Classical Games 4 Player Interface". CSDB. Retrieved 9 March 2011.

[7] "4 Player Interface". Protovision. Retrieved 6 March 2011.

[8] Ninja. "How to connect 8 snes-pads to your c64 v1.1". Hitmen. Retrieved 3 March 2011.

[9] "SNES-Pad-Tooldisk #1". CSDB. Retrieved 19 October 2011.

[10] Groepaz. "The Digital Excess & Hitmen 4-Player Joystick adapter". Hitmen. Retrieved 3 March 2011.

[11] "Eight joysticks on the C64". c128.net (from German via Google translate). Retrieved 6 August 2014.

[12] "C64.cz". Retrieved 26 August 2014.

[13] "Singular Crew 4Player Adapter for Commodore 64". Retrieved 27 August 2014.

[14] "zsibvasar.hu". Retrieved 27 August 2014.

[15] "What is NEW with the 2.3 release?". VICE team. Retrieved 4 March 2011.

[16] "VICE Feature Requeset tracker: Joystick emulation of Starbyte/Kingsoft adapters". SourceForge. Retrieved 15 March 2012.

[17] "Nostalgia releases". Retrieved 6 March 2011.

[18] Gutjahr, Christoph. "C64-games supporting four joysticks". Delirium BBS. Retrieved 3 March 2011.

[19] Synnes, Stig. "4-joystick adapter". Retrieved 6 March 2011.

[20] Da!NyL's workspace. Retrieved 20 October 2011.

[21] . GameBase 64 Collection. Retrieved 26 January 2012.

[22] . The GameBase 64 Collection. Retrieved 20 October 2011.

[23] DXS & HIT 4-Player Joystick Adapter. CSDB. Retrieved 20 October 2011.

[24] International Karate+ GOLD. CSDB. Retrieved 20 October 2011.

[25] Ninja. CSDB. Retrieved 17 February 2014.

[26] Rampage Gold. CSDB. Retrieved 20 October 2011.

[27] Grubz. Singular Crew. Retrieved 20 October 2011.

[28] Nostalgia releases. Retrieved 3 March 2011.

[29] Phong. CSDB. Retrieved 20 October 2011.

[30] Tour de France. CSDB. Retrieved 20 October 2011.

[31] Alone in the green. CSDB. Retrieved 20 October 20111.

[32] MULE. CSDB. Retrieved 7 December 2011.

[33] . P1X3L.net. Retrieved 26 January 2012.

[34] Square Attack. CSDB. Retrieved 20 October 2011.

[35] Icon Run. CSDB. Retrieved 8 June 2012.

[36] Bomberland. RGCD. Retrieved 2 April 2014.

[37] Octron. CSDB. Retrieved 6 August 2014.

[38] Schlimeisch Mania. CSDB. Retrieved 6 August 2014.

[39] Race. CSDB. Retrieved 11 March 2015.

2.15.7 External links

- Protovision 4 player interface page, shop and list of compatible games

- Hitmen building instruction on 4 player adapters and list of 4 player games

- C64.cz Inception 8 joysticks adapter interface, shop and list of compatible games

2.16 Commodore 64 peripherals

Commodore 64 Home Computer

This article is about the various external peripherals of the Commodore 64 home computer.

2.16.1 Storage

Tape drives

Main article: Datasette
In the United States, the 1541 floppy disk drive was

Commodore Datasette 1530

widespread. By contrast, in Europe, the C64 was often used with cassette tape drives (Datasette), which were much cheaper, but also much slower than floppy drives. The Datasette plugged into a proprietary edge connector on the Commodore 64's motherboard. Standard blank audio cassettes could be used in this drive. Data tapes could be write-protected in the same way as audio cassettes, by punching out a tab on the cassette's top edge.

The Datasette's speed was very slow (about 300 baud). Loading a large program at normal speed could take up to 30 minutes in extreme cases. Many European software developers wrote their own fast tape-loaders which replaced the internal KERNAL code in the C64 and offered loading times more comparable to disk drive speeds. **Novaload** was perhaps the most popular tape-loader used by British and American software developers. Early versions of **Novaload** had the ability to play music while a program loaded into memory, and was easily recognizable by its black border and digital bleeping sounds on loading. Other fast-loaders included load screens, displaying computer artwork while the program loaded. More advanced fast-loaders included minigames for the user to play while the program loaded from cassette. One such minigame fastloader was Invade-a-Load.

Users also had to contend with interference from magnetic fields. Also, not too dissimilar to floppy drive users, the Datasette's read head could become dirty or slip out of alignment. A small screwdriver could be used to align the tape heads, and a few companies capitalized by selling various commercial kits for Datasette head-alignment tuning.

As the Datasette lacked any random read-write access, users had to either wait while the tape ran its length, while the computer printed messages like "SEARCHING FOR ALIEN BOXING... FOUND AFO... FOUND SPACE INVADERS... FOUND PAC-MAN... FOUND ALIEN BOXING... LOADING..." or else rely on a tape counter number to find the starting location of programs on cassette. Tape counter speeds varied over different datasette units making recorded counter numbers unreliable on different hardware.

An optional streaming tape drive, based upon the QIC-02 format, was available for the Xetec Lt. Kernal hard drive subsystem (see below). They were expensive and few were ever sold.

A similar concept to the ZX Microdrive (85 kB) was the extremely fast "**Phonemark 8500 Quick Data Drive**" which has 128 kB capacity using a micro-cassette storage unit and used the C2N Datasette. The concept eventually succumbed to floppy drives.[1][2]

Backup to VHS tapes were offered by **DC Electronics** with their cartridge **WHIZZARD** in 1988. Which could handle

5.8 kB/s and included "freezer" capabilities.[3]

Floppy disk drives

Commodore 1541 Floppy Drive

Although usually not supplied with the machine, floppy disk drives of the 5¼ inch (1541, 1570 and 1571) and, later, 3½ inch (1581) variety were available from Commodore.

The 1541 was the standard floppy disk drive for the Commodore 64, with nearly all disk-based software programs released for the computer being distributed in the 1541 compatible floppy disk format. The 1541 was very slow in loading programs because of a poorly implemented serial bus, a legacy of the Commodore VIC-20.

The 1541 disk drive was notorious for not only its slow performance and large physical size compared to the C64 (the drive is almost as deep as the computer is wide), but also for the drive mechanisms installed during early production runs, which quickly gained a bad reputation for their mechanical unreliability.

Perhaps the most common failure involved the drive's read-write head mechanism losing its alignment. Due to lack of hardware support for detecting track zero position, Commodore DOS formatting routines and many complex software copy-protection schemes (which used data stored on nonstandard tracks on floppies) had to rely on moving the head specified number of steps in order to make sure that the desired head position for formatting or reading the data was reached. Since after physically reaching track zero, further movement attempts caused the head drive mechanism to slam (producing the infamous, loud, telltale knocking sound) into a mechanical stop, the repetitive strain often drove the head mechanism out of precise alignment, resulting in read errors and necessitating repairs. As a side note: some demos exploited the sound generated by the head moving stepper motor to force the disk drive to play crude tunes ("Bicycle Built For Two" was one) by varying the frequency of step requests sent to the motor.

Also, as with the C64, 1541 drives tended to overheat due

to a design that did not permit adequate cooling (potentially fixed by mounting a small fan to the case). Many of the 1541's design problems were eventually rectified in Commodore's 1541-II disk drive, which was compatible with the older units. The power supply unit was not housed inside the drive case; hence, the 1541-II size was significantly smaller and did not overheat.

Because of the drive's initial high cost (about as much as the computer itself) and target market of home computer users, BASIC's file commands defaulted to the tape drive (device 1). In order to load a file from a commercial disk, the following command must be entered:

LOAD "*",8,1

In this example, '*' designates the last program loaded, or the first program on the disk, '8' is the disk drive device number, and the '1' signifies that the file is to be loaded not to the standard memory address for BASIC programs, but to the address where its program header tells it to go—the address it was saved from. This last '1' usually signifies a machine language program.

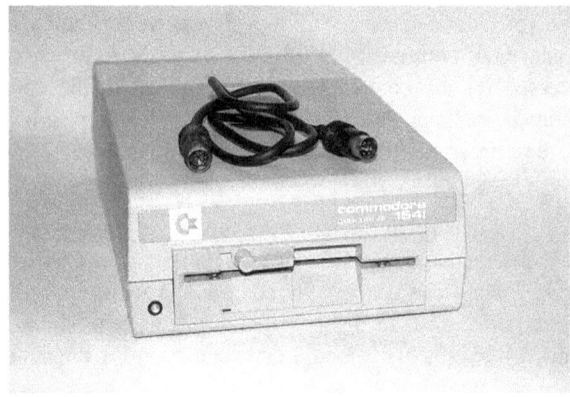

Commodore 1541C Floppy Drive, 2nd model

Commodore 1541-II Floppy Drive, 3rd model

Not long after the 1541's introduction, third-party developers demonstrated that performance could be improved with software that took over control of the serial bus signal lines and implemented a better transfer protocol between the computer and disk. In 1984 Epyx released its *FastLoad* cartridge for the C64, which replaced some of the 1541's slow routines with its own custom code, thus allowing users to load programs in a fraction of the time. Despite being incompatible with many programs' copy protection schemes, the cartridge became so popular among grateful C64 owners (likely the most-widespread third-party enhancement for the C64 of all time) that many Commodore dealers sold the Epyx cartridge as a standard item when selling a new C64 with the 1541.

As a free alternative to FastLoad cartridges, numerous pure software *turbo-loader* programs were also created that were loaded to RAM each time after the computer was reset. The best of these turbo-loaders were able to accelerate the time required for loading a program from the floppy drive by a factor of 20x, demonstrating the default bus implementation's inadequacy. As turbo-loader programs were relatively small, it was common to place one on almost each floppy disk so that it could be quickly loaded to RAM after restart.

The 1541 floppy drive contained a MOS 6502 processor acting as the drive controller, along with a built-in disk operating system (DOS) in ROM and a small amount of RAM, the latter primarily used for buffer space. Since this arrangement was, in effect, a specialized computer, it was possible to write custom controller routines and load them into the drive's RAM, thus making the drive work independently of the C64 machine. For example, certain back up software allowed users to make multiple disk copies directly between daisy-chained drives without a C64.

Several third party vendors sold an IEEE-488 general purpose interface bus adapter for the C64, which plugged into the machine's expansion port. Outside of BBS operators, few C64 owners took advantage of this arrangement and the accompanying IEEE devices that Commodore sold (such as the SFD-1001 1-megabyte 5¼ inch floppy disk drive, and the peripherals originally made for the IEEE equipped PET computers, such as the 4040 and 8050 drives and the 9060/9090 hard disk drives).

As an alternative to the feeble performing 1541 or the relatively expensive IEEE bus adapter and associated peripherals, a number of third-party serial-bus drives such as the MSD Super Disk and Indus GT appeared that often offered better reliability, higher performance, quieter operation, or simply a lower price than the 1541, although often at the expense of software compatibility due to the difficulty of reverse engineering the DOS built into the 1541's hardware (Commodore's IEEE-based drives faced the same issue due to the dependence of the DOS on features of the Commodore serial bus).

Like the IEEE-488 interface, the serial bus offered the ability to daisy chain hardware together. This led to Commodore producing (via a third party) the Commodore 4015, or VIC-switch. This device (now rarely seen) allowed up to 8 Commodore 64s to be connected to the device along with a string of peripherals, allowing each computer to share the connected hardware.

It was also possible, without requiring a VIC-switch, to connect two Commodore 64s to one 1541 floppy disk drive to simulate an elementary network, allowing the two computers to share data on a single disk (if the two computers made simultaneous requests, the 1541 admirably handled one whilst returning an error to the other, which surprised many people who expected the 1541's less-than-stellar drive controller to crash or hang). This functionality also worked with a mixed combination of PET, VIC-20, and other selected Commodore 8-bit computers.

In the mid-1980s, a 2.8-inch floppy disk drive, the Triton Disk Drive and Controller, was introduced by Radofin Electronics, Ltd. It was compatible with the Commodore 64 as well as other popular home computers of the time, thanks to an operating system stored on an EPROM on an external controller. It offered a capacity of 144/100 kilobytes non-formatted/formatted, and data transfer rates of up to 100 kilobytes per second. Up to 20 files could be kept on each side of the double-sided floppy disks.

Later in the 1990s, Creative Micro Designs produced several powerful floppy disk drives for the Commodore 64. These included the FD-Series serial bus compatible 3.5″ floppy drives (FD-2000, FD-4000), which were capable of emulating Commodore's 1581 3.5″ drive as well as implementing a native mode partitioning which allowed typical 3.5″ high-density floppy disks to hold 1.6 MB of data—more than MS-DOS's 1.4 MB format. The FD-4000 drive had the advantage of being able to read hard-to-find enhanced floppy disks and could be formatted to hold 3.2 MB of data. In addition, the FD series drives could partition floppy disks to emulate the 1541, 1571 and 1581 disk format (although unfortunately, not the emulated drive firmware), and a real time clock module could be mounted inside the drive to time-stamp files. Commercially, very little software was ever released on either 1581 disk format or CMD's native format. However, enthusiasts could use this drive to transfer data between typical PC MS-DOS and the Commodore with special software, such as SOGWAP's Big Blue Reader.

There was one other 3.5″ floppy drive available for the Commodore 64. The "TIB 001" was a 3.5″ floppy drive that connected to the Commodore 64 via the expansion port, meaning that these drives were very fast. The floppy disks themselves relied on an MS-DOS disk format, and being based on cartridge allowed the Commodore 64 to boot from them automatically at start-up. These devices

appeared from a company in the United Kingdom, but unfortunately did not become widespread due to non-existent third-party support. In an article in *Zzap!64* of November 1991, several software houses interviewed believed that the device came to the market too late to be worthy of supporting.

Hard drives

Seagate ST 506 5¼-inch HDD with cover removed.

Late in 1984, Fiscal Information Inc., of Florida, demonstrated the Lt. Kernal hard drive subsystem for the C64. The Lt. Kernal mated a 10 megabyte Seagate ST-412 hard drive to an OMTI SASI intelligent controller, creating a high speed bus interface to the C64's expansion port. Connection of the SASI bus to the C64 was accomplished with a custom designed host adapter. The Lt. Kernal shipped with a disk operation system (DOS) that, among other things, allowed execution of a program by simply typing its name and pressing the Return key. The DOS also included a keyed random access feature that made it possible for a skilled programmer to implement ISAM style databases.

By 1987, the manufacturing and distribution of the Lt. Kernal had been turned over to Xetec, Inc., who also introduced C128 compatibility (including support for CP/M). Standard drive size had been increased to 20 MB, with 40 MB available as an option, and the system bus was now the industry-standard small computer system interface, better known as SCSI (the direct descendant of SASI).

The Lt. Kernal was capable of a data transfer rate of over 38 kB per second (65 kB per second in C128 fast mode). An optional multiplexer allowed one Lt. Kernal drive to be shared by as many as sixteen C64s or C128s (in any combination), using a round-robin scheduling algorithm that took advantage of the SCSI bus protocol's ability to handle multiple initiators and targets. Thus the Lt. Kernal could be conveniently used in a multi-computer setup, something that was not possible with other C64-compatible hard drives.

Production of the Lt. Kernal ceased in 1991. Fortunately, most of the components used in the original design were industry standard parts, making it possible to make limited repairs to the units. In 2010, a re-creation of the Lt. Kernal was produced by MyTec Electronics. It was called the Rear Admiral HyperDrive and used an upgraded DOS called RA-DOS. The Rear Admiral parts could be used to upgrade the older Lt. Kernal, e.g. chips from the Rear Admiral host adapter could be used to upgrade the chips in the Lt. Kernal host adapter; or if the Lt. Kernal is missing its host adapter, the Rear Admiral host adapter could be used in its place.

Also available for the Commodore 64 was the Creative Micro Designs CMD HD-Series. Much like the Commodore 1541 floppy drive, the CMD HD could connect to the Commodore 64's serial bus, and could operate independently of the computer with the help of its on-board hardware. A CMD HD series drive included its own SCSI controller to operate its hard drive mechanism, in addition to hosting a battery powered real-time clock module for the time-stamping of files. The stock operating speeds of the CMD HD-Series units were not very much faster than the stock speeds of a 1541 floppy drive, but the units were fully JiffyDOS compatible. Faster parallel transfers were possible with the addition of another CMD product, the CMD RAMLink and a special parallel transfer cable. With this arrangement, the performance of the system doubled that of the Lt. Kernal. One advantage the CMD products had was software compatibility, especially with GEOS, that prior solutions lacked. CMD ultimately missed opportunities to develop any features for the drive's auxiliary port (such as a printer spooler feature promised in the CMD HD user manual). Support for external SCSI devices (such as CD-ROM and Zip drives) was also noticeably missing. SCSI devices could be connected and chained to the external SCSI port, but could not be used from the HD without workarounds or special software.

The ICT DataChief included a 20MB hard drive, along with an Indus GT floppy drive, along with a 135-watt power supply in a case designed to house an IBM PC Compatible computer.[4]

User operation of these hard drive subsystems was similar to that of Commodore's floppy drives, with the inclusion of special DOS features to make best use of the drive's capabilities and to effectively manage the vast increase in storage capacity (up to a maximum of 4GB). An unavoidable problem was that total 1541 compatibility could not be achieved, which often prevented the use of copy-protected software, software fastloaders, or any software whose operation depended on exact 1541 emulation.

The enthusiast-built "IDE64 interface" was designed late in the 1990s, attaching itself in the Commodore 64's expansion port, and allowing users to attach common IDE hard drives, CD-ROM and DVD drives, ZiP and LS-120 floppy drives to their Commodore 64s. Later revisions of the interface board provided an extra compact flash socket. The IDE interface's performance is comparable to the RAMLink in speed, but lacks the intelligence of SCSI. Its main advantage lies in being able to use inexpensive commodity hard drives instead of the more costly SCSI units. 1541 compatibility is not as good as commercially developed hard drive subsystems, but continues to improve with time.

In late 2011, MyTec Electronics developed and sold the Rear Admiral Thunderdrive, a clone of the CMD HD. Though using more modern components and a smaller form factor in comparison to the CMD HD, the Thunderdrive maintained full compatibility with the CMD HD.

2.16.2 Input/Output

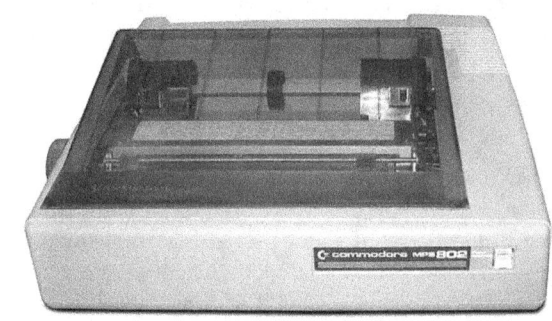

Commodore MPS 802

Printers

A number of printers were released for the Commodore 64, both by Commodore themselves and by third-party manufacturers.

Commodore-specific printers were attached to the C64 via the serial port and were capable of being daisy chained to the system with other serial port devices such as floppy drives. By convention, printers were addressed as device #4-5 on the Commodore serial bus.

Dot-matrix A series of dot-matrix printers were sold by Commodore, including the MPS 801 (OEM Seikosha GP 500 VC) and the MPS 803, although many other third-party printers like the Okimate 10 and Okidata 120 were popular too - some having more advanced printing features than any of Commodore's models. Most Commodore-branded printers were rebranded C. Itoh or Epson models with Commodore serial interface.

Daisy wheel   Commodore also produced the DPS-1101 daisy wheel printer, which produced letter quality print similar to a typewriter, and which typically cost more than the computer and floppy disk drive together. The MPS-1000 dot matrix printer was introduced along with the C-128.[5] Commodore 1526 is a rebranded MPS 802.[6][7]

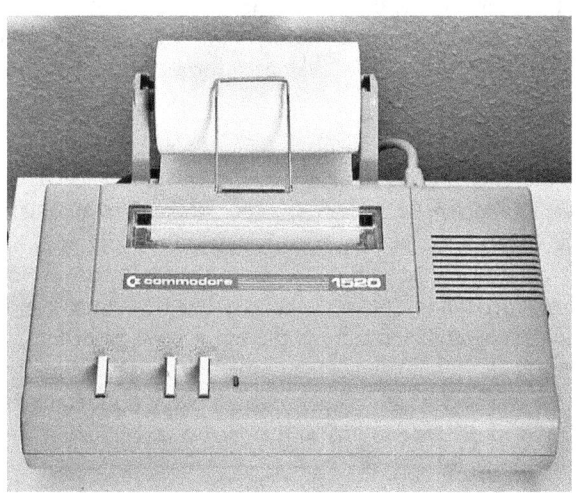

Commodore 1520 plotter

Plotter   A mini plotter device, the Commodore 1520, could plot graphics and print text in four colors by using tiny ballpoint pens.

The 1520 was based upon the Alps Electric DPG1302, a mechanism which also formed the basis of numerous other inexpensive plotters for home computers of the time (e.g. the Atari 1020).[8][9]

Third-party printer interfaces and buffers   Since there were severe shortcomings of early Commodore printers, CARDCO released the Card Print A (C/?A) printer interface that emulated Commodore printers by converting the Commodore-style IEEE-488 serial interface to a Centronics printer port to allow numerous 3rd-party printers to be connected to a Commodore 64, such as Epson, Okidata, C. Itoh.[10] A second model, a version that supported printer graphics was released called the Card Print +G (C/?+G), supported printing Commodore graphic characters using ESC/P escape codes. CARDCO released additional enhancements, including a model with RS-232 output, and shipped a total over 2 million printer interfaces. Xetec also released a series of printer interfaces. With a parallel interface, the QMS KISS laser printer, the most inexpensive then available at $1995, could be used.[11] Later,

CMD created the GeoCable which allowed PS2-type inkjet and laser printers to work under GEOS with a special device driver.

Printer buffer with 64 kB RAM for the IEC IEEE-488 serial bus existed too, like the "Brachman Associates Serial Box Print Buffer".[12]

Input devices

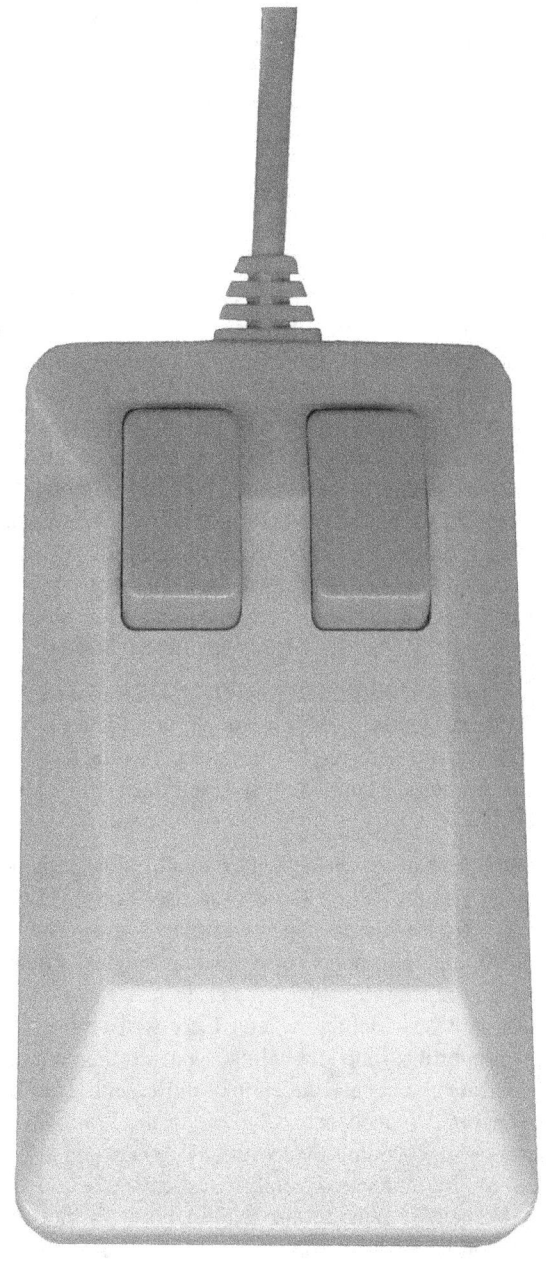

Commodore mouse

Commodore produced joystick controllers for the Com-

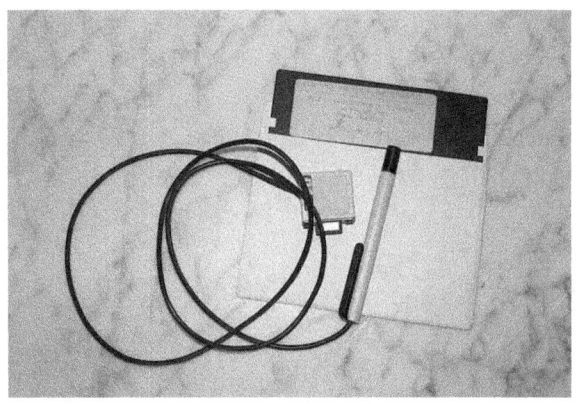

C64 Lightpen with its Software of the Company Rex-Datentechnik

modore 64, largely compatible with Atari joysticks, as well as paddles (which were not Atari compatible). Commodore's paddles were originally intended for the VIC-20, and few C64 games could take advantage of them. Commodore's joysticks were often derided because they were not particularly robust, especially for extreme gameplay. Many gaming enthusiasts preferred third-party joysticks, while some enthusiasts even built their own joysticks and controllers for the Commodore 64, or modified controllers from other systems to work on it. While the Commodore 64 only had two joystick ports for use, a few different kinds of joystick adapters were constructed by enthusiasts, which allowed up to four or eight joysticks to be used on the Commodore 64, with appropriate programming. Only about 20 games (by 2011) can take advantage of these however.

The "Atari CX85 Numerical Keypad" consists of a numeric keypad featuring the 17 keys [escape], [no], [delete], [yes], 0-9, [.], [-] and [+/enter].[13] It connects to the C64 joystick port using the Atari 2600 style interface with a DB9F plug.[14][15]

Commodore had two models of computer mouse, namely the 1350 and the 1351. These were used with GEOS as well as software such as Jane and Magic Desk. The earlier 1350 was only capable of emulating a digital joystick, by sending rapid 8 directional signals as it was moved, and thus was not very useful. The later 1351 used a more traditional proportional mode, sending signals to the computer that indicate amount and direction of movement. The 1351 also supported a mode identical to that of the 1350. CMD's SmartMouse was compatible with 1351-aware and also included a third button and a built in real-time clock module as well. The NEOS mouse also existed, but it was not compatible with 1351-aware software as it was simply a joystick emulator.

Several Companies produced Lightpens with its own drawing software for the Computer, e.g. the Inkwell light pen which was compatible with GEOS.

The Koala Pad graphics tablet was also available, came with its own paint software, and was compatible with GEOS as well. Suncom's Animation Station was another graphics tablet for the C64.[16]

Car positioning system

A senior test technicians at CGAD Productions operations developed and installed the CarPilot *Computerized Automotive Relative Performance Indicator and Location of Transit.* Which may be one of the first GPS type mapping systems to be tested, circa 1984. It utilizes a Commodore 64, 12V DC to 5V DC converter, video player/recorder, datasette, and a TV monitor.[17]

The monitor page 1 displays battery voltage, water temperature, engine oil pressure, fuel level, vehicle speed, engine rotation speed, lock/no-lock condition of the automatic transmission torque converter, and on/off condition of the air conditioning clutch. All except the last two were incorporated with a "buzzer" alarm system that indicate malfunction. Another feature is the one-second-precision 24-hour clock. Estimated arrival time with 1s precision, distance traveled which is incremented every 80 meters and estimated distance to arrival that is also decremented with same value, 80 meters.[17]

Page 2 displayed the vehicle position along the map. Vehicle location indication is calculated from distance traveled. The accuracy of the vehicle location is dependent of the digital map construction and the accuracy of the local map used to construct the digital map. The best hope for accuracy is 800 m. But accuracy of one car length in 35 km has been realized. The use assembly language was necessitated to keep up with sensor input. One advantage with the system is the ability to create one's own digital maps and thus eliminate the need to buy such ones for every trip. The software to accomplish this task was written in Basic.[17]

Robotics

With *computing*, *robot trainer*, and *plotter-scanner*, Fischertechnik rose as the first manufacturer of modular building blocks into the computer age. Interfaces for all popular home computers at the time were made, including Apple II, Commodore 64 and Acorn, and later for Schneider, Atari ST and IBM PC. Programming languages to drive the models included GW-BASIC, Turbo Pascal and in the later kits (1991) an in-house programming tool Lucky Logic.

The "Commocoffee 64" is an espresso maker controlled by the C64[18] in 1985.[19]

See also: List of educational programming lan-

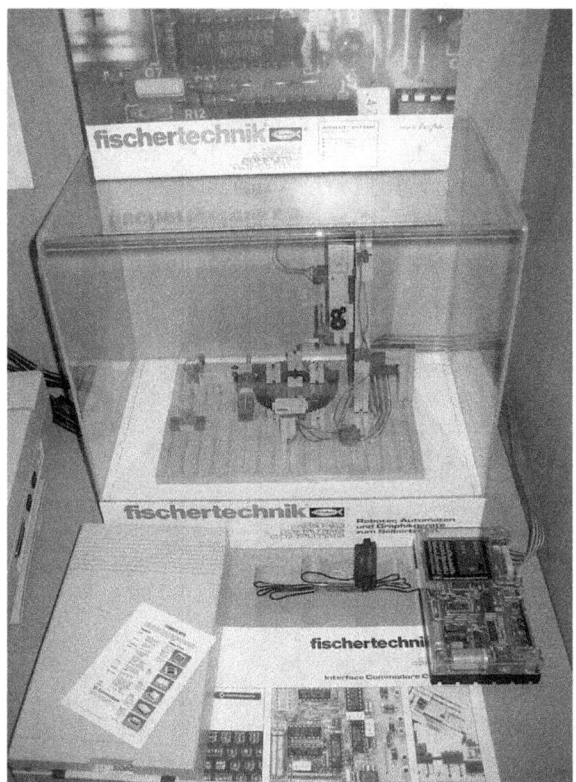

Fischertechnik computing *with a C64 interface*

guages

Relay controller

The Handic "VIC REL" controller provides protected input and output using 6 relay outputs and 2 optocoupler inputs. The output relays are capable of 24 V / 10 W and the inputs respond to 5-12 V DC. The device also provides (+5V) and (−5V) at 50 mA to activate inputs. The device is programmed on the VIC-20 with *POKE 37138,63* and I/O at *37136*. And on C64 with *POKE 56579,63* and I/O at *56577*. The intended applications were burglar alarms, garage doors, door locks, heating elements, lamps, transmitters, remote controllers, valves, pumps, telephones, accumulators, irrigation systems, electrical tools, stop watches, ventilators, humidifiers, etc.[20]

Analog to digital converters

There are audio Analog-to-digital converters (A/D) like the "A/D Wandler (DELA 87393)" based on 8-bit ADC0809 chip for the C64/128[21] with a maximum sampling frequency of 10 kHz.[22] and the Sound Ultimate Xpander 6400 (SUX 6400) based on the 8-bit ADC0804 chip with

a maximum sampling frequency of 11 kHz. Plain sound digitizers like "Sound Digitizer (REX 9614)" that converts analog sound into 2-bit samples.[21] The latter could also be accomplished using the Datasette and software tricks.[23][24]

Biofeedback EEG/EMG

In 1987 there was a cartridge port device to measure EEG directly for use in exercise programs, called "BodyLink" produced by the company Bodylog in New York, USA.[25] Schippers-Medizintechnik in Germany produced a user port attached EMG device to allow a physician to analyze such things as stress level, and assisting in finding a better position for work.[26]

Handscanners

The "Scanntronik Handyscanner 64" is a hand held scanner that uses the C64 user port.[1][27]

Frame grabbers

Frame grabbers like the "PAL Colour Digitizer" that connect via the user port, will turn an analog composite video frame into a digital picture on the C64.[1] The "Print Technik Video Digitizer" connects via the user port and uses CVBS video signal that has to be still for 4 seconds in order to be sampled and can then be saved either as 320×200 monocolour or 160×200 multicolour (4 colours).[28]

Video generator

80 column mode could be used by installing the "BI-80" cartridge released 1984[29] from "Batteries Included" which is built around the 6545 video chip. It includes an expansion ROM that adds BASIC 4.0 commands. One can control which 40/80 column mode is active by software. On power up, the 40-column mode is active.[30][31]

Another 80 column card using the cartridge port was the "DATA20 XL80" introduced in 1984[32] Costing 400 000 Lira in 1985.[33]

The "Z80 Video Pack 80" enabled black and white 80-column screen and CP/M using a Zilog Z80.[15]

Teletext

To download pages and software transmitted via the teletext broadcast system. The UK company "Microtext" provided their "Teletext adaptor" and tuner that interfaced with the

TV-aerial and the C64/128 user port. Software was provided on a C-10 tape.[1][34] Which were priced at 114.80 GBP inc. p/p in 1987.[35]

2.16.3 Communication

Modems

Commodore VIC Modem

As Commodore offered a number of inexpensive modems for the C64, such as the 1650, 1660, 1670, the machine also helped popularize the use of modems for telecommunications.[36][37] The 1650 and 1660 were 300 Baud, and the 1670 was 1200 baud. The 1650 could only dial Pulse. The 1660 had no sound chip of its own to generate Touch Tones, so a cable from the monitor /audio out was required to be connected to the 1660 so it could use the C64 sound chip to generate Touch Tones. The 1670 used a modified set of Hayes AT commands.

This modem is required for Medical Manager for EDI operations.

The Commodore 1650 shipped with a rudimentary piece of terminal software called Common Sense. It provided basic Xmodem functionality and contained a 700 line scrollback feature.

In the United States, Commodore offered the Commodore Information Network, a CompuServe SIG devoted to its products and users. Later, Quantum Computer Services (which became America Online) offered an online service called Quantum Link for the C64 that featured chat, downloads, and online games. In the UK, Compunet was a very popular online service for C64 users (requiring special Compunet modems) from 1984 to the early 1990s. In Australia, Telecom (now Telstra) ran an online service called Viatel and sold modems for the C64 for use with the service. In Germany the very restrictive rules of the state-owned

telephone system prevented widespread use of inexpensive, non-telco licensed modems, prompting the use of inferior acoustic couplers instead. Access to Bildschirmtext, the state-owned telco's own dial-up online service, was possible via special add-on hardware like the Commodore "BTX Decoder Modul" [38] or the Commodore "BTX Decoder Modul II".[1][39][40]

Radio communication

"Microlog AIR-1 Radio Interface Cartridge" that use the cartridge port with builtin ROM software for RTTY and morse code communications.[41]

"RTTY-CW Interface C-64" uses the User port for RTTY communications.[42]

"Auerswald ACC-64" longwave time signal receiver.[43]

RS-232 port

Like the VIC-20, the C64 lacked a real UART chip such as the 6551 and used software emulation. This limited the maximum speed to an error-prone 2400 bit/s. Third-party cartridges with UART chips offered better performance.

Later in the Commodore 64's life, CMD developed two serial communications cartridges for Commodore Computers, the "Swiftlink" (1990[44] - 38 400 bit/s)[45] and the "Turbo 232" (1997[46] - 230 400 bit/s).[47] The latter was capable of handling a 56k Hayes modem reliably at full speed on a Commodore 64, enabling reasonable dial-up internet access speeds.

The Retro-Replay expansion cartridge enabled the addition of the **Silver Surfer** add-on serial board, which also enabled 56k modem connections, and the **RR-Net** add-on serial board, which allows for broadband internet access, as well as LAN.

Also, on November 5, 2005 Quantum Link Reloaded was launched enabling C64 enthusiasts to experience all the features of the original Quantum Link service in present-day with some enhancements for free.

IEEE-488

The Commodore 64 IEEE-488 Cartridges were made by various companies, but Commodore themselves never made one for the Commodore 64/128 family. One of uses were harddiscs like the Commodore D9060.

Some other interfaces without pictures available:

- E-LINK Serial to IEEE Interface. (contains 65C02, 6522 and 4 kB ROM)

- Buscard II Interface. (contains a 6532, 6821 (PIA) and 8 kB ROM, and a 256 byte PROM)

- INTERPOD - An Interface box, that converts IEEE-488 to CBM (IEC) serial & RS-232 serial.

2.16.4 Other peripherals

Commodore 1702 video monitor

The Commodore 1701 and 1702 were 13-inch (33 cm) color monitors for the C64 which accepted as input either composite video or separate chrominance and luminance signals, similar to the S-Video standard, for superior performance with the C64 (or other devices capable of outputting a separated signal). Other monitors available included the 1802 and 1902. Introduced in 1986, the 1802 featured separate chroma and luma signals, as well as a composite green screen mode suitable for the C-128's 80 column screen.[48] The 1902 had a true RGBI 80-column mode compatible with IBM PCs.

Early in the Commodore 64's life, Commodore released several niche hardware enhancements for sound manipulation. These included the "Sound Expander", "Sound Sampler", "Music Maker" overlay, and External music keyboard. The Sound Expander and Sound Sampler were both expansion cartridges, but had limited use. The Sound Sampler in particular could only record close to two seconds of audio, rendering it largely useless. The Music Maker was a plastic overlay for the Commodore 64 "breadbox" keyboard, which included plastic piano keys corresponding to keys on the keyboard. The External keyboard was an add-on which plugged into the Sound Expander. These hardware devices did not sell well, perhaps due to their cost, lack of adequate software, marketing as home consumer de-

vices, and an end result that turned many serious musicians off.

Possibly the most complex C64 peripheral was the Mimic Systems Spartan, which added an entire new computer architecture to the C64, with its own 6502 CPU and expansion bus, for software and hardware compatibility with the Apple II series. Announced shortly after the Commodore 64 itself at a time when little software was available for the machine, the Spartan did not begin shipping until 1986, by which time the C64 had acquired an extensive software library of its own.[49] Essentially an Apple II+ compatible computer that used the 64's keyboard, video output, joysticks, and cassette recorder, the Spartan included 64kB RAM, a motherboard with a 6502 CPU on a card, 8 Apple-compatible expansion slots, an Apple-compatible disk controller card, and a DOS board to add to your 1541 disk drive. The DOS board was optional, but if it was not installed an Apple Disk II or compatible drive would be required to load software. The long delay between announcement and availability, along with heavy promotion including full-page ads running monthly in the Commodore press, made the Spartan an infamous example of vaporware.

Gamesware produced a gaming peripheral for the Commodore 64 in 1988, where a target board was attached to the computer using the RS-232 port to enable use of its *Gamma Strike* suite of games.

CMD produced a SID symphony cartridge later in the Commodore's life. A reworking of the original Dr. T's SID Symphony cartridge, this cartridge gave the Commodore another SID chip for use to play stereo SID music. This saved Commodore 64 users from needing to modify their computer motherboards to enable it with dual SID chips.

Creative Micro Designs (CMD) was the longest-running third-party hardware vendor for the Commodore 64 and Commodore 128, hailed by some enthusiasts as being better at supporting the Commodore 64 than Commodore themselves. Their first commercial product for the C64 was a KERNAL based fast loader and utility chip called JiffyDOS. It was not the first KERNAL-based enhancement for the C64 (SpeedDOS and DolphinDOS also existed), but was perhaps the best implemented. The benefits of a KERNAL upgrade meant that the cartridge port was free for use (which would have normally been taken up by an Epyx FastLoad cartridge or an Action Replay), however the downside meant that one had to manually remove computer chips from the C64's motherboard and associated floppy drives to install it. Aside from the usual 1541 fast load routines, JiffyDOS contained an easy to use DOS and a few other useful utilities.

RAM expansions

Over the years, a number of RAM expansion cartridges were developed for the Commodore 64 and 128. Commodore officially produced several models of RAM expansion cartridges, referred to collectively as the 17xx-series Commodore REUs. While these devices came in 128, 256, or 512 kB sizes, third-party modifications were quickly developed that could extend these devices to 2 MB, although some such modifications could be unstable. Some companies also offered services to professionally upgrade these devices.

Typically, most Commodore 64 users did not require a RAM expansion. Very little of the available software was programmed to make use of expansion memory. The cost of the units (and the requirement to add a heavy-duty power supply) also was a factor in the limited usage of RAM expansion cartridges. The volatility of DRAM was also a factor in the limited usage, as the RAM expansion cartridges were normally used for fast RAM disk storage, data stored on them would be lost at any power failure.

Aside from power-supply problems, the other main downfall of the RAM expansions were their limited usability due to their technical implementation. The RAM in the expansion cartridges was only accessible via a handful of hardware registers, rather than being CPU-addressable memory. This meant that users could not access this RAM without complicated programming techniques. Furthermore, simply adding the RAM expansion did not provide any kind of on-board RAM disk functionality (though a utility disk was supplied with some REUs, which provided a loadable RAM disk driver).

One popular exception to the disuse of the REUs was GEOS. As GEOS made heavy use of a primitive, software-controlled form of swap space, it tended to be slow when used exclusively with floppy disks or hard drives. With the addition of an REU, along with a small software driver, GEOS would use the expanded memory in place of its usual swap space, increasing GEOS' operating speed.

Due to the lack of available 17xx-series Commodore REUs, and then their later discontinuation, Berkeley Softworks, the publishers of GEOS, developed their own 512 kB RAM expansion cartridge - the GeoRAM. This device was purposely designed for use with GEOS, although some REU-aware programs were later adapted to be able to use it. Some time later, the GeoRAM was cloned by another company to form the BBGRAM device (which also sported a battery backup unit). The GeoRAM used a banked-memory design where portions of the external SRAM were banked into the Commodore 64's CPU address space. This method provided substantially slower transfer speeds than the single-cycle-per-byte transfer speeds of the Com-

modore REUs. A benefit of using SRAMs was lower power consumption which did not require upgrading the Commodore 64's power supply.

Eventually the Super 1750 Clone, a third-party clone of Commodore's RAM expansions was developed, designed in such a way as to eliminate the need for a heavy-duty power supply.

PPI devised their own externally powered 1 or 2 MB RAM expansion, marketed as the PPI/CMD RAMDrive, which was explicitly designed to be used as a RAM disk. Its primary feature was that the external power supply kept the formatting and contents of the RAM safe and valid while the computer was turned off, in addition to powering the device in any case. A driver was provided on the included utilities disk to allow GEOS to use the RAMdrive as a regular 'disk' drive.

CMD later followed up with the RAMLink. This device operated similar to the RAMDrive, but could address up to 16 MB of RAM in the form of a 17xx-series REU, Geo-RAM, and/or an internal memory card, which also provided a battery-backed realtime clock for file time/date stamping of files saved to it. It also features a battery backup, thus preserving the RAM's contents. Drivers were provided with the RAMLink to allow GEOS to use its memory as either a replacement for swap space, or as a regular 'disk' drive.

CMD's Super CPU Accelerator came after this, and could house up to 16 MB of direct, CPU-addressable RAM. Unfortunately, there was no on-board or disk-based RAM disk functionality offered, nor could any existing software make use of the directly addressable nature of the RAM. The exception is that drivers were included with the unit to explicitly allow GEOS to use that RAM as a replacement for swap space, or as a regular 'disk' drive, as well as to make use of the acceleration offered by the unit.

EPROM programmers

Programmers for EPROMs like 2716 - 27256 using common programming voltages (Vpp) of 12.5, 21, and 25 V were available by connecting a device to the user port of the C64.[50] These devices could cost 100 USD in 1985. The device often included a zero insertion force (ZIF) socket and a LED indicating when the EPROM chip was being programmed.[51] The cartridge port was also used by some programmer devices.[21]

Freezer, Reset, and Utility cartridges

Probably the most well-known hacker and development tools for the Commodore 64 included "Reset" and "Freezer" cartridges. As the C64 had no built-in soft re-

Micro Maxi Prommer, EPROM burner for C64 user port

set switch, reset cartridges were popular for entering game "POKEs" (codes which changed parts of a game's code in order to cheat) from popular Commodore computer magazines. Freezer cartridges had the capability to not only manually reset the machine, but also to dump the contents of the computer's memory and send the output to disk or tape. In addition, these cartridges had tools for editing game sprites, machine language monitors, floppy fast loaders, and other development tools. Freezer cartridges were not without controversy however. Despite containing many powerful tools for the programmer, they were also accused of aiding software pirates to defeat software copy protections. Perhaps the best known freezer cartridges were the Datel "Action Replay", Evesham Micros Freeze Frame MK III B, Trilogic "Expert", "The Final Cartridge III", and Super Snapshot cartridges.

The Lt. Kernal hard drive subsystem included a push button on the host adapter called ICQUB (pronounced "ice cube"), which could be used to halt a running program and capture a RAM image to disk. This would work with most copy-protected software that did not do disk overlays and/or bypass the KERNAL ROM jump table. The RAM image was runnable only on the Lt. Kernal system on which it was captured, thus preventing the process from being used to pirate software.

Music and Synthesizer utilities

As the Commodore 64 featured a digitally controlled semi-analogue synthesizer as its sound processor, it was not surprising to discover an abundance of software and hardware designed to expand upon its capabilities.

Various assemblers, notators, sequencers, MIDI editing and mixer automation software were created which allowed users and programmers to create or record musical pieces

of impressive technical complexity. Some software of note has included the Kawasaki Synthesizer range, Music System notation and MIDI suite, the MIDI-compatible Instant Music 'idiot-proof' sequential composer, and the Steinberg Pro-16 MIDI sequencer, the precursor to Cubase.

Notable hardware included various brands of MIDI cartridges, plug-in keyboards (such as the Color Tone or the Sound Chaser 64), Commodore's own SFX range which included a sound sampler and Sound Expander plug-in synthesizer and keyboard, the more recent Commodulator oscillator wheel and the Prophet 64 sequencer and synthesizer utility cartridge. The Passport Designs MIDI Interface is said to be one of the best designs and had the most software supported model available.[15]

Recently a few professional musicians have used the Commodore 64's unique sound to provide some or all of the synthesizer parts required for their performances or recordings; an example being the band Instant Remedy. Also noteworthy is the Commodore 64 Orchestra who specialize in rearranging and performing music originally composed and coded for the Commodore 64 games market. Its patron is celebrated Commodore composer Rob Hubbard.

Apple II+ emulation box

The Mimic Systems "Mimic Spartan Apple II+ compatibility box" enabled C64 users to run Apple II+ software.[52] It came with the "DOS Card" addition, an Apple II disk controller that was installed inside the Commodore 1541 disk drive, between the floppy logic board and the drive mechanism. In normal mode the circuit simply passed signals through but at the flick of a switch it could take over the mechanism and turn the drive into an Apple II drive. The potential for grave damage to both Apple II and 1541 floppies was enormous and often happened. The box had 24 jumpers to configure. Applesoft BASIC was included and very compatible, since it was created by disassembling the binary from the Applesoft ROM and reordering the assembly level instructions such that the binary image would be different. One could set up various debugging and use slave computing to enable fast 3D rendering etc. The box had functionality to switch video between C64 and Apple. The second advertisement were put into the COMPUTE!'s Gazette in 1986.[53]

CP/M with Z80 CPU cartridge

The Commodore C64 CP/M Cartridge used the C1541 floppy drive that was incapable to read any existing CP/M disk format. The cartridge were equipped with a Zilog Z80 CPU running at circa 3 MHz.[54] On the C128 the INT and IORQ signals are used such that the Z80 can make use of interrupts.[55]

CPU accelerators

Like the Apple II family, third-party acceleration units providing a faster CPU appeared late in the C64's life. Due to timing issues with the VIC-II chip - the same issues that caused the 1540 disk drive to be incompatible and the 128's "fast mode" to be 80 column-only - CPU accelerators for the 64 were much more complex and expensive to implement than for other computers. So while accelerators based on the WDC 65C02, usually running at 4 MHz, and on the 65816 at up to 20 MHz appeared, they appeared too late and were too expensive to gain widespread use.

The first CPU accelerator seen was called the "Turbo Process" by a Bonn, Germany, based company called Roßmöller GmbH. It used a Western Design Center 65816 running at 4.09 MHz. Code ran from faster static RAM on the accelerator expansion port cartridge. As the VIC chip can only see the internal DRAM memory, writes had to be mirrored to the internal memory, write cycles would slow the operation of the processor to accomplish this.

The *Turbo Master CPU*, produced by US based Schnedler Systems, was a blue expansion port device which clocked in at 4.09 MHz. It also had a JiffyDOS option. It was a copy of the Turbo Process system. Early Turbo Process circuit boards shipped with PAL chips that did not have their security fuses blown, this made copying the design quite easy. The Turbo Master CPU had one beneficial modification, the bit to toggle the high-speed mode on was "0" in memory location $00 as opposed to the "1" the Turbo Process. A lot of software would write zeros to this location turning off the high-speed mode on the Turbo Process - this was considered a design flaw that was fixed by the Turbo Master. No known litigation took place over the copying of the German company's design.

The most well-known accelerator for the C64 is probably Creative Micro Designs' SuperCPU, which gives the C64 a 20 MHz processor (instead of ~1 MHz) and up to 16 MB of RAM if combined with CMD's *SuperRam-Card*. Understandably, due to a very limited "market" and number of developers, there has not been much software tailored for the SuperCPU to date— however GEOS was supported. Among the few offerings available include the GEOS-compatible operating system, Wheels; a Wheels-based web browser called "The Wave", a Unix/QNX-like graphical OS called Wings, some demos, various classic games modified for use with the SuperCPU, and a shooter game in the old *Katakis*-style called *Metal Dust*.

Present and Future devices

While CMD no longer produces Commodore hardware, new peripherals are still being developed and produced, mostly for mass storage or networking purposes.

The MMC64 cartridge allows the C64 to access MMC- and SD flash memory cards. And several revisions and add-ons have been developed for it to take advantage of extra features. It features an Amiga clock port for connecting a RR-Net Ethernet-Interface, an MP3 player add-on called 'mp3@c64' has even been produced for it.

In February 2008, Individual Computers started shipping the MMC Replay. It unites the MMC64 and the Retro Replay in one cartridge, finally built with proper case-fit in mind (even including the RRnet2 Ethernet add-on). It contains many improvements, such as C128 compatibility, a built-in .d64 mounter (not speedloader-compatible though, because the 1541 CPU is not emulated), 512 kB ROM for a total of eight cartridges, 512 kB RAM, a built-in flash-tool for cartridge images and wider support for various types of cartridges (not merely Action-replay-based).

In April 2008, the first batch of *1541 Ultimate* shipped, a project by the hobbyist VHDL-developer Gideon Zweijtzer. This is a cartridge that carries an Action Replay and Final Cartridge (whatever the user prefers) and a very compatible FPGA-emulated 1541 drive that is fed from a built-in SD-card slot (.d64, prg etc.). The difference to other SD-based and .d64 mounting cartridges like the MMC64, Super Snapshot 2007 or MMC Replay is, that the 6502 that powers the 1541 Floppy and the 1541's mechanical behavior (even sound) is fully emulated, making it theoretically compatible with almost anything. Fileselection and management is done via a third button on the cartridge that brings up a new menu on screen. The 1541 Ultimate also works in standalone mode without a c-64, functioning just like a normal Commodore 1541 would. Disk-selection of .d64s is then done via buttons on the cartridge, power is supplied via USB. There is a "Plus-Version" available with an extra 32 Megabytes of RAM (as REU and for future use), the basic version has just enough RAM for the advertised functions to work. In October 2008, the second and third batch of 1541 Ultimates were produced to match the public demand for the device. The regular version without the 32MB RAM was dropped since there was no demand for it. Due to public demand there is also a version with Ethernet now. In 2010 a completely new PCB and software has been developed by Gideon Zweijtzer to facilitate the brand new 1541-Ultimate-II cartridge.

The IDE64 interface cartridge provides access to parallel ATA drives like hard disks, CD/DVD drives, LS-120, Zip drives, and CompactFlash cards. It also supports network drives (PCLink) to directly access a host system over various connection methods including X1541, RS-232, Ethernet and USB. The operating system called IDEDOS provides CBM/CMD compatible interface to programs on all devices. The main filesystem is called CFS, but there's read-only support for ISO 9660 and FAT12/16/32. Additional features include BASIC extension, DOS Wedge, file manager, machine code monitor, fast loader, BIOS setup screen.

Today's computer mice can be attached via the Micromys interface that can process even optical mice and similar. There are also various interfaces for plugging the 64 to a PC keyboard.

A special board for converting Commodore 64 video signals to standard VGA monitor output is also currently under development. Also a board to convert the Commodore 128's 80 column RGBI CGA-compatible video signal to VGA format was developed in late 2011. The board, named the C128 Video DAC, had a limited production run and was used in conjunction with the more widespread GBS-8220 board.

In September 2008, Individual Computers announced the Chameleon, a Cartridge for the Expansion Port that adds a lot of previously unseen functionality. It has a Retro-Replay compatible Freezer and MMC/SD-Slot, 16 MB REU and a PS/2 connector for a PC Keyboard. Support for a network adapter and battery-backed real time clock exists. The cartridge does not even have to be plugged into a Commodore 64 and can be used as a standalone device using USB power. Since the cartridge essentially also includes a Commodore One it is possible to include a VGA Port that outputs the picture to a standard PC monitor. The Commodore One core also allows the cartridge to be used as a CPU accelerator, and a core to run a Commodore Amiga environment in standalone mode also exists. Unlike most other modern day C64 hardware, this cartridge actually ships with a bright yellow case. Shipping was announced for Q1/2009, and currently the cartridge is available, although the firmware is in a beta state. A standalone mode docking station is under development.

Retro Innovations is shipping the *uIEC*[56] device, which utilizes the core design of the *SD2IEC* project to provide a mass media solution for Commodore 8-bit systems that utilize the Commodore IEC Serial Bus. NKCElectronics of Florida is shipping SD2IEC hardware which uses the sd2iec firmware. Manosoft sells the C64SD Infinity, another SD card media solution which uses the sd2iec firmware.

In Summer of 2013, another commercial variant of the SD2IEC-Device appears on market, the SD2IEC-evo2 from 16xEight.[57] This device uses an bigger uC (ATmega1284P) and has some extras such as Battery backed-up RTC, connector for LC-Display, Multicolour Status-LED, and so on already on board.

2.16.5 Notes

1. ^ Many users came to dread the telltale "RAT-AT-AT-AT-AT" knocking noise, since such knocking contributed to eventual disk drive alignment failure.

2. ^ A modification could be made to older model Commodore 64 motherboards to piggy-back a secondary SID sound chip to the original SID chip. The resulting modification enabled the Commodore 64 to play sound in 6-channel stereo with the appropriate software.

3. ^ The Commodore 64 had documented cartridge port pins which could be crossed to achieve a reset. In an attempt to activate game "reset" and various cheats, a large number of Commodore 64 users attempted to reset their machines by manually touching these pins 1 and 3 with wire while the computer was switched on. Many users made mistakes and missed the correct pins, blowing their C64's fuse and resulting in a costly repair. This achievement was later known as the "Hamster Reset" in "Commodore Format" magazine. Some users soldered these pins to a button, which they mounted in the C64's case for handy resetting. Some programs utilized reset protection (by having the string 'CBM80' [58] at $8000 in the memory) which could be worked around by shorting pins 1-3-9 the same way as the "Hamster Reset" pin 9 (on the top side as opposed to pins 1 & 3 on the bottom) being the EXROM ROM expansion pin (thus overwriting data at $8000–$9fff).

2.16.6 See also

- Computers: Commodore 64, VIC-20

- Floppy Drives: Commodore 1541, 1551, 1570, 1571, 1581

- Commodore 64 disk / tape emulation

2.16.7 References

[1] "Hardware". bithunter.siz.hu. 2012-01-30. Retrieved 2013-06-21.

[2] "coll_quick_data_drive.jpg". bithunter.siz.hu. 2012-01-23. Retrieved 2013-06-21.

[3] "tt". web.tiscali.it. 2012-09-22. Retrieved 2013-06-17.

[4] "RUN Magazine issue 40".

[5] "Run Issue 30 Jun 1986".

[6] "Chronology of Commodore Computer History, Jack Tramiel". 090505 commodore.ca

[7] "Here be Commodore Computers. Be in Awe.". 090505 zimmers.net

[8] "What are the Atari 1020, 1025, 1027, and 1029 Printers?". *faqs.org (Atari 8-Bit Computers: Frequently Asked Questions section)*. Retrieved 2015-03-22. = Commodore 1520 / Oric MCP40 / Tandy/Radio Shack CGP-115 /..; made by ALPS [..] 20, 40 and 80-column modes

[9] "The Texas Instruments HX-1000 Printer/Plotter Photos". *Hexbus.com*. Other printer plotters that use variants of the ALPS DPG1302 plotter mechanism include the: Commodore 1520, Tandy CGP-115, Sharp CE-150, Atari 1020, Mattel Aquarius 4615

[10] "CARDCO Card Print A (C/?A) - Printer Interface For The Commodore 64 and VIC-20". *COMPUTE Magazine* (34): 251. March 1983.

[11] "RUN Magazine issue 36".

[12] "commodore.ca | Rare Commodore Computer Hardware Picture / Photo Gallery". commodore.ca. 2012-12-11. Retrieved 2013-06-21.

[13] "Commodore%2064_128%20Key%20Pad_Atari.jpg". commodore.ca. 2011-03-29. Retrieved 2013-06-21.

[14] "Review: Atari CX85 Numerical Keypad". atarimagazines.com. May 1983. Retrieved 2013-06-21.

[15] "Products | Commodore 64 History, Manuals & Photo's 64C 64GS". commodore.ca. 2011-03-30. Retrieved 2013-06-21.

[16] Infoworld Media Group, Inc (1984-07-09). *Software for the Suncom Graphics Tablet*.

[17] "commodore-64-car-pilot.jpg". commodore.ca. 2011-03-30. Retrieved 2013-06-21.

[18] "The Commocoffee-64 » Coolest Gadgets". coolest-gadgets.com. Retrieved 2013-06-21.

[19] "commocoffee-commodore-64-coffee-maker.jpg". commodore.ca. 2011-03-29. Retrieved 2013-06-21.

[20] "VIC REL" (PDF). bombjack.org. 2009-11-14. Retrieved 2013-06-21.

[21] "- Rex Datentechnik - Retroport". retroport.de. 2013-06-14. Retrieved 2013-06-21.

[22] "ADC0808/ADC0809 8-Bit µP Compatible A/D Converters with 8-Channel Multiplexer" (PDF). learn-c.com. 2010-04-15. Retrieved 2013-06-21.

[23] "Could the Datasette players play music cassette tapes too? - Commodore 64 (C64) Forum". lemon64.com. Retrieved 2013-06-21.

[24] "C64 Tape player - Commodore 64 (C64) Forum". lemon64.com. Retrieved 2013-06-21. 5 poke53265,0 10 for i=0 to 25:read a:poke49152+i,a:next:sys49152 90 data 120,165,1,41,223,133,1,162,0,160,15,169,16 91 data 44,13,220,240,251,142,24,212,140,24,212,208,243

[25] COMPUTE!'s GAZETTE, January 1987, Issue 43, Vol. 5, No. 1 |page=10

[26] "The C64 as a medical aid". mos6502.com. 2012-09-21. Retrieved 2013-07-06.

[27] "coll_handyscanner.jpg". bithunter.siz.hu. 2012-01-23. Retrieved 2013-06-21.

[28] "coll_pal.jpg". bithunter.siz.hu. 2012-01-23. Retrieved 2013-06-21.

[29] "B I - 8 0 80 Column Display by Batteries Included" (PDF). mikenaberezny.com. Retrieved 2013-06-17.

[30] "BI-80 Display Adapter". mikenaberezny.com. 2012-01-28. Retrieved 2013-06-17.

[31] "coll_bi-80.jpg". bithunter.siz.hu. 2012-01-23. Retrieved 2013-06-21.

[32] "B80.jpg". web.tiscali.it. 2012-09-16. Retrieved 2013-06-17.

[33] "Data 20 Corporation XL 80 video a 80 colonne per C 64" (PDF). digitanto.it. 2010-02-13. Retrieved 2013-06-17.

[34] "coll_microtext.jpg". bithunter.siz.hu. 2012-01-23. Retrieved 2013-06-21.

[35] Your Commodore, Issue 35, August 1987, page 7

[36] http://www.zimmers.net/cbmpics/ouser1.html

[37] http://archive.org/stream/VIC-1600_VICMODEM_ 1982_Commodore/VIC-1600_VICMODEM_1982_ Commodore_djvu.txt

[38] "- Hardware B-C - Retroport". retroport.de. 2013-06-14. Retrieved 2013-06-21.

[39] "coll_btx.jpg". bithunter.siz.hu. 2012-01-23. Retrieved 2013-06-21.

[40] "Bildschirmtext-Museum: Hardware-Btx-Decoder: Meine Sammlung". btxmuseum.de. Retrieved 2013-06-21.

[41] "coll_microlog_air-1.jpg". bithunter.siz.hu. 2012-01-23. Retrieved 2013-06-21.

[42] "empty". bithunter.siz.hu. 2012-01-23. Retrieved 2013-06-21.

[43] "coll_acc64.jpg". bithunter.siz.hu. 2012-01-23. Retrieved 2013-06-21.

[44] "Mike Naberezny – CMD SwiftLink RS-232". mikenaberezny.com. 2012-01-28. Retrieved 2013-06-17.

[45] "USR Modem - comp.sys.cbm | Google Groups". groups.google.com. 1996-08-06. Retrieved 2013-06-17.

[46] "File:Turbo232 top.jpg - ReplayResources". ar.c64.org. 2010-06-15. Retrieved 2013-06-17.

[47] "CMD Turbo232 High speed modem interface" (PDF). ar.c64.org. 2010-06-15. Retrieved 2013-06-17.

[48] "Commodore 1802 User's Manual".

[49] "RUN Magazine issue 36 December 1986".

[50] "empty" (PDF). bombjack.org. 2009-01-26. Retrieved 2013-06-21.

[51] "EPROM Programmers handbook for the C64 and C128" (PDF). bombjack.org. 2009-01-02. Retrieved 2013-06-21.

[52] "VC&G | [Retro Scan of the Week] Apple II Box for C64". vintagecomputing.com. 2013-03-25. Retrieved 2013-06-21.

[53] "Mimic Systems' Spartan | Applefritter". applefritter.com. 2013-06-21. Retrieved 2013-06-21.

[54] "Commodore 64 CP/M Cartridge". devili.iki.fi. 2006-02-24. Retrieved 2013-06-21.

[55] "Ruud's Commodore Site: C/PM-cartridge for the C64". baltissen.org. 2009-07-30. Retrieved 2013-06-21.

[56] Retro Innovations - uIEC

[57] SD2IEC-evo2

[58] The string 'CBM80' being represented by the hex bytes C3 C2 CD 38 30

2.16.8   External links

- Individual Computers - Makers of MMC64 and RR-series products

- 16xEight Digital Retrovation - Makers of innovative new hardware for Commodore 8-Bit Computers

- Protovision - Makers of various new hardware upgrades

- Lemon64 - Includes some of the best Commodore 64 music software

- Home Recording - Music discussion board thread linking to many others relevant to C64 music

- RUN Magazine Issue 39 May, 1986 special printer issue

- elektronik.si: Vic-Rel internal PCB

Manuals

Commodore

- Commodore VIC-1541 Floppy Drive: User Manual, Technical Reference

- Commodore VIC-1515 Printer: User Manual

- Commodore VIC-1525 Printer: User Manual

CARDCO

- CARDCO Card Print A (C/?A) Printer Interface: User Manual, Addendum

- CARDCO Card Print +G (C/?+G) Printer Interface: User Manual, Supplement

2.17   Commodore 65

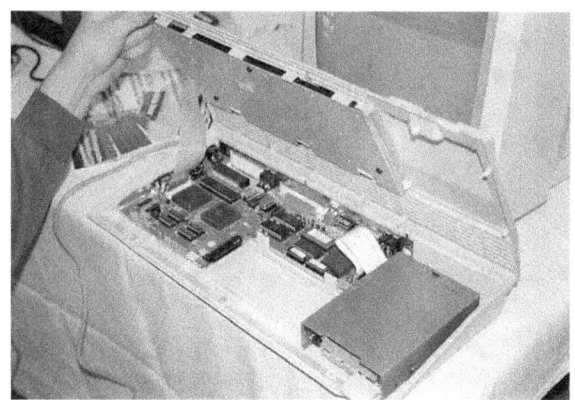

A Commodore 65 prototype opened up, revealing its internal disk drive

The **Commodore 65** (also known as the **C64DX**) is a prototype computer created at Commodore Business Machines in 1990-1991. It is an improved version of the Commodore 64, and it was meant to be backwards-compatible with the older computer, while still providing a number of advanced features close to those of the Amiga.

2.17.1   History

In September 1989 *Compute!'s Gazette* noted that "Sales of the 64 have diminished rapidly, Nintendo has eaten big holes in the market, and the life of the old warhorse computer should somehow be extended." Noting that Apple had developed the IIGS to extend the life of its Apple II series, the magazine asked "Will Commodore take the same tack?", then continued:[1]

CSG 4510 ("Victor")

F011B (floppy disk controller)

CSG 4567 ("Bill")

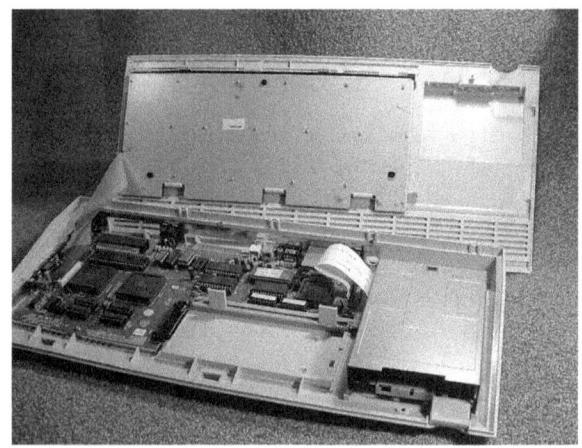

Opened chassis

"Elmer" and "Igor" (programmable logic)

Motherboard inscription

The latest rumor says *Yes*. We've heard reports from several sources of a new machine from Commodore—A 64GS, if you will. This machine is reportedly driven by a GE802, a version of the 65816 microprocessor (which is a 16-bit version of the 6502 chip), and runs at 4 MHz (by comparison, the 64 runs at 1 MHz; the Amiga, at slightly over 7 MHz). It comes with 128K of

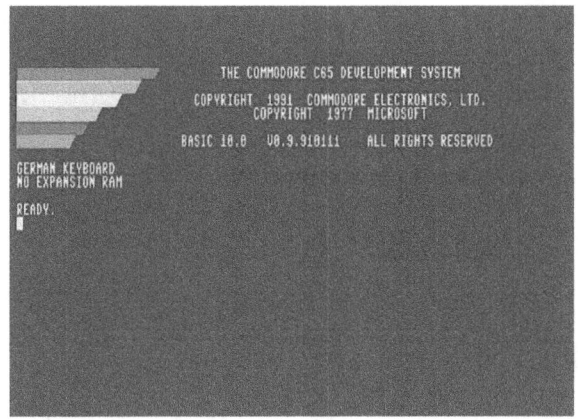

Start screen

RAM and is expandable to one megabyte. Fully expanded, it supports 256 colors. Maximum resolution is a stunning 640 X 400 pixels. We've also heard that it has a 64 mode so that 64 owners can purchase a much more powerful machine and still use their software library. The 64GS reportedly comes with a built-in 3 1/2-inch disk drive and will support the 1581. But, our sources say, it does not support the 1541 or the 1571 drive (uh, excuse me, please pass the bologna). All we've heard about sound in the new machine is that it's "enhanced" and features stereo output. The final tidbit is that the 64GS will retail in the $300-$350 range when it debuts in November.

The *Gazette* added, "Our sources also report that there is a great deal of infighting at Commodore as to whether the machine should be released. The sales staff wants to get the machine out the door, while the naysaying engineers have dubbed it 'son of Plus/4.'"[1] While the next issue reported that "the latest rumor is that such a machine will never see the light of day",[2] Fred Bowen and others at Commodore in 1990–1991 developed the Commodore 65 (C65) as a successor to the C64. In the end of 1990 the decision to create the C65 was taken.[3] The project was cancelled later on.

When Commodore International was liquidated in 1994, a number of prototypes were sold on the open market, and thus a few people actually own a Commodore 65. Estimates as to the actual number of machines found on the open market range from 50 to 2000 units.[4] As the C65 project was cancelled, the final 8-bit offering from CBM remained the triple-mode, 1–2 MHz, 128 kB (expandable), C64-compatible Commodore 128 of 1985.

In April 22, 2015 © MEGA - Museum of Electronic Games & Art[5] announced a recreation of this computer featuring similar specifications and technologies. Also backwards

compatible with the Commodore 64,[6] the Mega 65[7][8] features Commodore 65 like hardware[6] and is compatible with newer technologies such as HDMI.[9] The recreation of the Commodore 65 computer will be released in the third quarter of 2016.[6]

2.17.2   Technical specifications

- The CPU named CSG 4510 R3 is a custom CSG[10] 65CE02 (a MOS 6502 derivative), combined with two MOS 6526 complex interface adapters (CIAs)

- 3.54 MHz clock frequency (the C64 ran at 1 MHz)

- A new VIC-III graphics chip named CSG 4567 R5, capable of producing 256 colors from a palette of 4096 colors; available modes include 320×200×256, 640×200×256, 640×400×16, 1280×200×16, and 1280×400×4 (X×Y×color depth, i.e. number of colors/bit planes)

  - Supports all video modes of VIC-II

  - Textmode with 40/80 × 25 characters

  - Synchronizable with external video source (genlock)

  - Integrated DMA controller (bit blit)

- Two CSG 8580R5 SID sound chips producing stereo sound (the C64 has one SID)

  - Separate control (left / right) for volume, filter and modulation

- 128 kB RAM, expandable to 8 MB using a RAM expansion port similar to that of the Commodore Amiga 500

- 128 kB ROM

- Heavily improved BASIC: Commodore BASIC 10.0 (the C64 has the relatively feature-weak BASIC 2.0, which was almost 10 years old by this time.)

- One internal 3½" DSDD floppy disk drive

- Keyboard with 77 keys and an inverted T directional cursor block

Different views

Ports

Left side:

- Power +5V DC at 2.2A and +12V DC at 0.85A[11]

- 2× Control ports DE9M[11]

Back:

- Expansion port 50-pin[11]

- Serial bus using 6-pin DIN for 1541/1571/1581[11]

- User port: parallel 24-pin (without 9V AC)[11]

- Stereo 2× RCA connector[12] for left and right channel[11]

- RGBA video DE9F[11][13]

- RF video[11]

- Composite video 8-pin DIN[11]

- External fast floppy drive port - mini-DIN-8[11]

Bottom flap:

- RAM expansion[11]

Dimensions: ~ 46 cm wide, 20 cm deep, 5.1 cm high[11]

Chipset names

The custom chips of the C65 were not meant to have names like the custom chips in the Amiga. Although there are names printed near the chip sockets on various revisions of the circuit board, they were not intended as names for the chips. According to former Commodore engineer Bill Gardei, "The Legend on the PCB was to let others in the organization know [whom] to go to for advice on the chips. We did have an issue with that. But that wasn't the name of the chip at the time. The 4567 was always called the VIC-3. I can see why others outside of Commodore made the connection. But again—no—we never called these chips 'Victor' or 'Bill'."[14]

The custom chips for the C65 are:

- CSG 4510: processor (commonly called "Victor" after Victor Andrade)

- CSG 4567: VIC-III graphics processor (commonly called "Bill" after Bill Gardei)

- CSG 4151: DMAgic DMA controller (designed by Paul Lassa)

- F011C: FDC (floppy disk controller, also designed by Bill Gardei)

The C65 also contains one or two programmable logic arrays depending on the version:

- ELMER: PAL16L8 (C65 versions 1.1, 2A, 2B), PAL20L8 (C65 versions 3-5)

- IGOR: PAL16L8 (C65 version 2B only)

DOS

In contrast to previous 8-bit computers from Commodore, the C65 has a complete DOS through which the built-in 3.5" floppy disk drive can be controlled. Disks used by the C65 have a storage capacity of 880kB and the drive is compatible with C1581. Since this format was uncommon for the former C64 owners, the C65 retains the serial IEC port for external Commodore disk drives. It's possible to use a 1541, 1571, 1581, or other similar model.

The DOS itself is based on the Commodore PET IEEE 8250 drive DOS. Since it can only deal with two floppy disk drives, including the internal, only one external drive may be connected to the internal floppy disk controller. Like earlier systems, up to 4 drives can be daisy-chained on the IEC port.

Interfaces

The C65 includes the same ports of the C64. In addition, there is a DMA port for memory expansion. The latter is attached just like on the Amiga 500 via a flap in the bottom of the bottom of the board.[15] The built-in floppy disk drive is connected in parallel, serial Commodore drives can be connected via the usual IEC port. A plug for a genlock was also provided. Only the port for datasette the C64 is no longer available, and the user port missing—like the Aldi C64—the 9 volt AC line. The expansion port differs significantly from all prior C64 variants and rather resembles that of C16.

2.17.3 Sales

In December 2009, a working C65 on the online auction site eBay achieved a sales price of €6060.[16][17] A computer with missing parts was in October 2011 sold for about 20100 USD. In April 2013 an eBay auction reached the highest auction price for an C65 at 17827 EUR.[18] The latest eBay auction from February 2015 closed at 20050 EUR.[19]

2.17.4 Notes

[1] Elko, Lance (1989-09). "Editor's Notes". *Compute's Gazette*. p. 4. Retrieved 4 March 2015. Check date values in: |date= (help)

[2] Elko, Lance (1989-10). "Editor's Notes". *Compute's Gazette*. p. 2. Retrieved 4 March 2015. Check date values in: |date= (help)

[3] "OLD-COMPUTERS.COM museum ~ Commodore C65". old-computers.com. Retrieved 2013-06-20.

[4] "Secret Weapons of Commodore: The Commodore 65". floodgap.com. 2007-07-01. Retrieved 2013-02-26.

[5] "MEGA | MEGA - Museum of Electronic Games & Art". *www.m-e-g-a.org*. Retrieved 2015-10-06.

[6] "MEGA65 Computer". *mega65.org*. Retrieved 2015-10-06.

[7] "Introducing the MEGA65 (8-bit) computer | MEGA - Museum of Electronic Games & Art". *www.m-e-g-a.org*. Retrieved 2015-10-06.

[8] "MEGA 65: Commodore 65 remake gets a physical release • /r/c64". *reddit*. Retrieved 2015-10-06.

[9] "Making a C64/C65 compatible computer in an FPGA". *c65gs.blogspot.com*. Retrieved 2015-10-06.

[10] Commodore Semiconductor Group, previously known as MOS Technology, Inc

[11] "C64DX System specification". zimmers.net. 2009-08-18. Retrieved 2013-06-21.

[12] "c65_html_1410c60e.gif". retrocommodore.com. 2012-12-16. Retrieved 2013-06-21.

[13] "commodore.ca/gallery/hardware/c65.jpg". commodore.ca. 2011-03-29. Retrieved 2013-06-21.

[14] "The Story Behind 'Bill' and 'Victor'". collectorcomputers.com. 2013-11-16. Retrieved 2013-11-16.

[15] "File:C65-open.jpg". commons.wikimedia.org. 1997. Retrieved 2013-06-20.

[16] "C65-Auktion auf eBay". cgi.ebay.at. Retrieved 2010-01-03. (dead link)

[17] "C65-Auktion auf eBay". img5.imagebanana.com. 2012-11-10. Retrieved 2010-01-04.

[18] "Sehr seltener Prototyp Commodore C65 aka C64DX aka C90". ebay.de. 2013-04-14. Retrieved 2013-04-18.

[19] "Ultra rare Commodore 65 / C65 / DX64 prototype, working, serial #22". ebay.fr. 2015-02-15. Retrieved 2015-02-15.

2.17.5 References

- On the Edge: The Spectacular Rise and Fall of Commodore (2005), Variant Press. ISBN 0-9738649-0-7.

2.17.6 External links

- Hi65: a high-level Commodore 65 emulator

- C65 page at 'The Secret Weapons of Commodore' website By Cameron Kaiser and The Commodore Knowledge Base

- FTP directory for the C65 at ftp.zimmers.net

- Andre Kaesmacher's C64DX Development Site

- C64DX System Specification document

- C65 System ROMs and Utility Software

- Commodore 65: Like The C64, But It's One Louder

- old-computers.com: LD-COMPUTERS.COM museum ~ Commodore C65, article on C65

- 8-Bit-Nirvana: Commodore 65

- heimcomputer.de: Commodore C65 Prototyp, German C65-site with many photos and info

- toxic-waste.de: Commodore C65 Information Page by TXW

- cbmmuseum.kuto.de: CCOM - Commodore 65

2.18 Commodore Datasette

The **Commodore 1530 (C2N) Datasette** (a portmanteau of *data* and *cassette*), was Commodore's dedicated magnetic tape data storage device. Using compact cassettes as the storage medium, it provided inexpensive storage to Commodore's 8-bit home/personal computers, notably the PET, VIC-20, and C64. A physically similar model **Commodore 1531** was made for the Commodore 16 and Plus/4 series computers.

2.18.1 Description and history

The Datasette contained built-in analog to digital converters and audio filters to convert the computer's digital information into analog sound and vice versa (much like a modem does over a telephone line). Connection to the computer was done via a proprietary edge connector (Commodore 1530) or mini-DIN connector (Commodore 1531). The absence of recordable audio signals on this interface made the Datasette and its few clones the only cassette recorders usable with CBM's machines, until aftermarket converters made the use of ordinary recorders possible.

The Datasette was more popular outside than inside the United States. U.S. Gold, which imported American computer games to Britain, often had to wait until they were converted from disk because most British Commodore 64 owners used tape.[1][2] *Computer Gaming World* reported in 1986 that British cassette-based software had failed in the United States because "97% of the Commodore systems in the USA have disk drives";[3] by contrast, MicroProse reported in 1987 that 80% of its 100,000 sales of *Gunship* in the UK were on cassette.[4] In the United States disk drives quickly became standard, despite the Commodore 1541 floppy drive costing roughly five times as much as a Datasette. In most parts of Europe, the Datasette was the medium of choice for several years after its launch, although floppy disk drives were generally available. The inexpensive and widely available audio cassettes made the Datasette a good choice for the budget-aware home computer mass market.

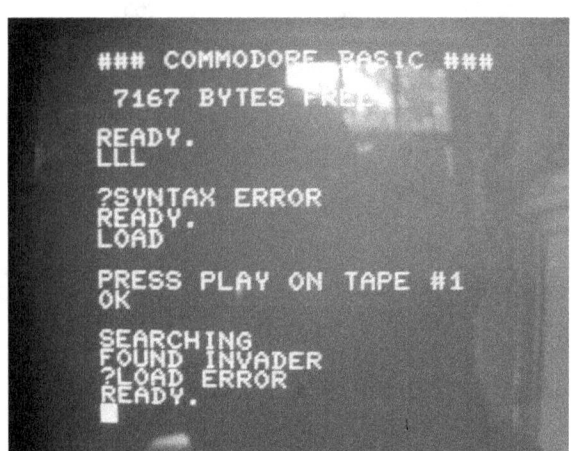

The Datasette loading process

The Datasette was slow albeit extremely reliable,[5][6] transferring data at around 50 bytes per second; even the very slow 1541 was significantly faster. Some years after the Datasette's launch, however, special *turbo tape* software appeared, providing much faster tape operation (loading and saving). Such software was integrated into most commercial prerecorded applications (mostly games), as well as being available separately for loading and saving the users' homemade programs and data. These programs were only widely used in Europe, as the US market had long since moved onto disks.

Datasettes could typically store about 100 kByte per 30 minute side.[7] The use of *turbo tape* and other fast loaders increased this number to roughly 1000 kByte.

2.18.2 Interface

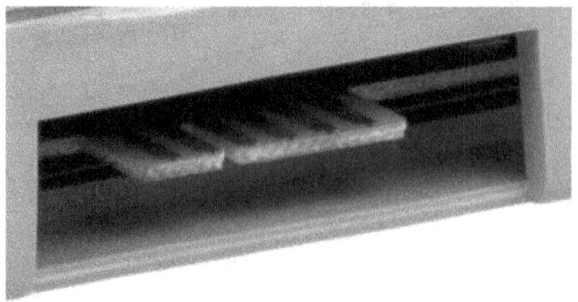

Commodore 64 cassette port

The Datasette has only one connection cable with a PCB edge connector at the computer end. All input/output signals to the datasette are all digital and so all digital to analog and vice versa is handled within the unit. Power is also included in this cable. The pinout is ground, +5 V DC, motor, read, write, key-sense.[8] The sense signal monitors the play, rewind, and fast-forward buttons, but cannot differentiate between them. A mechanical interlock prevented any two of them being pressed at the same time. Unregulated 6.36 V DC[9] is used to power the cassette motor.[10]

2.18.3 Physical coding

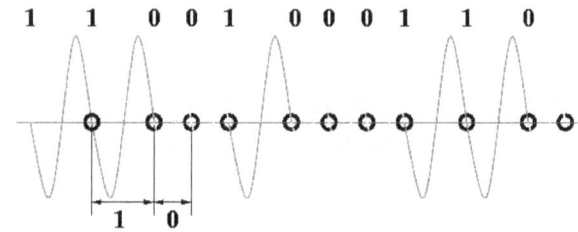

The resulting waveform from storing data

To record physical data, the zero-crossing from positive to negative voltage of the analog signal is measured. The resulting time between these positive to negative crossings is then compared to a threshold to determine whether the time since the last crossing is short (0) or long (1).[11] Note the lower amplitude for the shorter periods.

A circuit in the tape unit transforms the analog signal into a logical one or zero, which is then transmitted to the computer via the tape connector. Inside the computer, the first Complex Interface Adapter (6526) in the C64 senses when the signal goes from one to zero. This event is called trigger and causes an interrupt request. This event can be handled by a handler code, or simply discovered by testing bit 4 of location $DC0D. The points that trigger this event are indicated by the black circles in the figure.[11]

Inside the tape device the read head signal is fed into an operational amplifier (1) whose output signal is DC-filtered. Op-amp (2) amplifies and feeds an RC-filter. Op-amp (3) amplifies the signal again followed by another DC-filter. Op-amp (4) amplifies the signal into clipping the sine formed signal. The positive and negative rails for all op-amps are wired to +5V DC and GND. The clipped signal therefore fits into the TTL electrical level window of the schmitt trigger step that in turn feeds the digital cassette port.[12]

On the PAL version of the C64, the time granularity is 1.014 μs (for NTSC 0.978 μs). For a 300 bit/s data rate and where each bit uses 3284 clock cycles this means 3284 * 1.014 μs = 3330 μs/bit.

Once the bits can be decoded, they are fed into a shift register and are continuously compared to a special bit sequence. This bit sequence can also be seen as a byte. A bit-sequence match means that the stream is byte-synchronized. The first byte to compare with is called *lead-in byte*. If matched, it's compared to the *sync byte* as well.[11]

An example: Turbo Tape 64 has a *lead-in byte* $02 (binary 00000010), *sync byte* $09 (binary 00001001) and a following sync sequence of $08, $07, $06, $05, $03, $02, $01.[11]

2.18.4 Practical handling

Typical labeling of cassette inlays with the meter reading of the tape drive and the appropriate computer game titles

The way to arrange a "directory" and find software on tapes was accomplished by using an inlay sheet with counter position noted, and program name next to it.

The physical shape of the Commodore Datasette 1530/31 is a weight of 0.7 kg and measurement 19.5 cm wide, 5 cm high and 15 cm deep.

2.18.5 Main models

Used with the PET, VIC-20, C64/128

There are at least four main models of the 1530/C2N Datassette:

- The built-in Datassette in the original PET 2001: black cassette lid, five white keys, no tape counter, no SAVE LED

- Black body original shape model, black cassette lid, five black keys, no tape counter, no SAVE LED

- White body original shape model, black cassette lid, five black keys, with tape counter, no SAVE LED

- White body new shape model, silver cassette lid, six black keys, with tape counter and a red SAVE LED

The first two external models were made as PET peripherals, and styled after the PET 2001 built-in tape drive. The latter two were styled and marketed for the VIC-20 and C64. All 1530s were compatible with all those computers, as well as the C128.

In addition to this, some models came with a small hole above the keys, to allow access to the adjustment screw of the tape head azimuth position. A small screwdriver could thus easily be used to effect the adjustment without disassembling the Datassette's chassis.

Confusingly, the Datassette at various times was sold both as the *C2N DATASETTE UNIT Model 1530* and as the *1530 DATASSETTE UNIT Model C2N*. Note the difference in spelling (one *S* versus two) used on the original product packaging.[13]

Used with the C16/116 and Plus/4

Similar in physical appearance to the 1530/C2N models is the **Commodore 1531**, made for the Commodore 16 and Plus/4 series computers. This had a Mini-DIN connector in place of the PCB edge connector. This could be used with a C64/128 via an adaptor, which was supplied by Commodore with some units.

- Black/Charcoal body new shape model, silver cassette lid, six light gray keys, with tape counter and a red SAVE LED

2.18.6 Models

- The second, most common version of the 1530 C2N Datassette

- Datassette 1531

- One of the few clones

2.18.7 See also

- Magnetic tape data storage

- Fast loader

- IBM cassette tape

- Kansas City standard

2.18.8 References

[1] Anderson, Chris (June 1985). "On top of the US Goldmine". *Zzap!64* (interview). pp. 46–48. Retrieved 26 October 2013.

[2] Pountain, Dick (January 1985). "The Amstrad CPC 464". *BYTE*. p. 401. Retrieved 27 October 2013.

[3] Wagner, Roy (August 1986). "The Commodore Key". *Computer Gaming World*. p. 28.

[4] Brooks, M. Evan (November 1987). "Titans of the Computer Gaming World / MicroProse". *Computer Gaming World*. p. 16.

[5] "How TurboTape Works".

[6] "The Official Book for the Commodore 128".

[7] "Basic Commodore information".

[8] pinouts.ru - C64 Cassette pinout, 2012-01-15

[9] "250469 rev.A right". 100610 zimmers.net

[10] "250469 rev.A left". 100610 zimmers.net

[11] "How Commodore tapes work". 091205 wavprg.sourceforge.net

[12] Datasette service manual model C2N/1530/1531, preliminary, Oct. 1984 PN-314002-02

[13] Bo Zimmerman. "Faster than a speeding South American Grima Slug". *Commodore Gallery*. Retrieved 20 April 2012.

2.18.9 External links

- Similar Commodore tape drives

- Datasette photos

- Description of tape format with conversion utilities and code

- C2N232 project to build a hardware adaptor/software program to archive Commodore Datasette files to a modern computer.

- DC2N Homepage Digital C2N replacement project.

- Sketchup model of the Commodore Datasette 1530. Sketchup model of the Commodore Datasette 1530.

2.19 Commodore Educator 64

The **Educator 64**, also known as the **PET 64** and **Model 4064**, was a microcomputer made by Commodore Business Machines in 1983. It was sold to schools as a replacement for aging Commodore PET systems. Schools were reluctant to adopt the Commodore 64 "breadbox" design due to theft or vandalism of the smaller, more exposed components. The 4064 designation followed in line with the PET's 4008, 4016 and 4032 models as a 64kB 40-column model.

The innards of the Educator 64 were refurbished Commodore 64 motherboards and monochromatic green monitors. The area above the keyboard contained a quick reference card for BASIC 2.0 and Commodore DOS commands. The only differences between the Educator 64 and the other 64 models were the graphics capabilities, the built-in speaker, the sound amplifier with volume control, the 1/8-inch mini-jack for mono sound output to headphones, the internal power supply, and the keyboard which is missing the color abbreviations imprinted on the front edge of the number keys. The Educator 64 retained the ability to display shades of green, while the PET 64 and 4064 were monochrome-only. Though the PET 4008/4016/4032 computers had cases made entirely of metal, only the Educator 64's base was metal—the upper case was made of thick plastic.

The Educator 64 was not sold in great numbers. It suffered from its monochrome display - many 64 titles assumed the availability of color. And, by that time, the US education market was firmly in Apple's grasp.

2.19.1 External links

- The C64 is Schoolbound

- Secret Weapons of Commodore

- Commodore EDUCATOR 64 Model 4064

2.20 Commodore MAX Machine

The **Commodore MAX Machine**, also known as **Ultimax** in the United States and **VC-10** in Germany, was a home computer designed and sold by Commodore International in Japan, beginning in early 1982, a predecessor to the popular Commodore 64. The Commodore 64 manual mentions the machine by name, suggesting that Commodore intended to sell the machine internationally; however, it is unclear whether the machine was ever actually sold outside of Japan. It is considered a rarity.

Software was loaded from plug-in cartridges and the unit had a membrane keyboard and 2.0 KiB of RAM internally and 0.5 KiB of color RAM (1024*4bit). It used a television set for a display. It used the same chipset and 6510 CPU as the Commodore 64, the same SID sound chip, and compatible ROM cartridge architecture so that MAX cartridges will work in the C-64. The MAX compatibility mode in C-64 was later frequently used for "freezer" cartridges (such as the Action Replay), as a convenient way to take control of the currently running program.[1] It was possible to use a tape drive for storage, but it lacked the serial and user ports necessary to connect a disk drive, printer, or modem.

MAX Machine, accessories and retail packaging.

It was intended to sell for around 200 USD. Although the MAX had better graphics and sound capability, Commodore's own VIC-20, which sold for around the same amount of money, was much more expandable, had a much larger software library, and had a better keyboard—all of which made it more attractive to consumers.

Unlike the C-64, the MAX never sold well and was quickly discontinued.

2.20.1   See also

- Commodore 64

- Commodore 64 Games System

2.20.2   References

[1] "The Ultimax/Max Machine, The 64GS, The 64CGS". *The Secret Weapons of Commodore.* 2007-07-01. Retrieved 2008-08-10.

2.20.3   External links

- Page dedicated to the MAX Machine

- The UltiMax machine (a.k.a. VIC-10)

- The MAX Machine, the odd one out

2.21   Commodore OS

Commodore OS (full name: **Commodore OS Vision**) was a free-to-download Linux distribution developed by Commodore USA and intended for its PCs. The distribution was based on Debian and Linux Mint, available only for x86-64 architectures, and used the GNOME 2 desktop environment. The first public beta version was released on 11 November 2011.[2] It has been continually updated through Commodore OS Vision 0.8 Beta and never came out of beta phase.

Commodore OS Vision was a free download and the software was under continual development. There are no warranties regarding its usage or applicability. This operating system is no longer in development. The company is now closed and its web site is no longer active.

2.21.1   Compatibility

Commodore OS was not compatible with Commodore 64 software. It did contain VICE, an open-source program which emulates the Commodore 64, Commodore 128, CBM-II, Commodore PET, Commodore VIC-20 and Commodore Plus/4.[3]

2.21.2 Design

Commodore OS was designed as a way to imitate the look and feel of Commodore's legacy systems, and as a complement to the all-in-one-keyboard style of the personal computer. Commodore OS includes a large collection of specifically selected software aimed at creativity, gaming, media, and other tasks.

However, it was fully compatible only to CommodoreUSA products often causing kernel panic on general PCs. Improved Fusion version was promised but never released.

2.21.3 References

[1] "Commodore OS Vision Downloads". Commodore USA, LLC. Retrieved 16 July 2012.

[2] "Commodore OS Vision". DistroWatch. 2011-11-11. Retrieved 2011-11-12.

[3] "The VICE emulator". Viceteam. Retrieved 20 September 2012.

2.22 Commodore REU

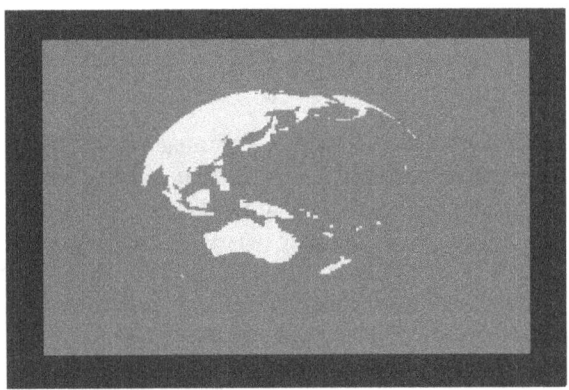

Official performance demonstration

Commodore's **RAM Expansion Unit (REU)** range of external RAM add-ons for their Commodore 64/128 home computers was announced at the same time as the C128. The REUs came in three models, initially the **1700** (128 KB) and **1750** (512 kB), and later the **1764** (256 kB, for the C64).

The need for the REU came about when Commodore management decided to not use the final version of the custom Memory Management Unit (MMU) which then limited the size of memory in spite of early discussion of a larger memory map. Engineers traveling to the 1985 CES show were confronted with flyers and billboards advertising a memory size that was no longer supported and finally the most upper management asked where the additional memory (Up to 512K) would plug in.

By the time of the 1985 CES show in Chicago, the engineers were able to display a spinning globe of the earth as a demonstration of Direct Memory Access (DMA) by the new REU units.

The REU hardware was designed by Frank Palia and the dedicated Integrated Circuit (IC) was designed by Victor Andrade. Fred Bowen and Terry Ryan adapted the kernel and Basic to accommodate the REU natively and Hedley Davis wrote the globe spinning demo which was an impressive display of animation in the mid 1980s.

2.22.1 Hardware description

Although the C128 could access more than 64 kB of RAM through bank switching, the memory inside the REU could only be accessed by memory-transfers (STORE/LOAD/SWAP/COMPAREs) between the main memory and the REU memory, thus, giving an equivalent to a (slow) small memory window. Additionally, the C128's built-in BASIC 7.0 had three statements, STASH, FETCH, and SWAP, for storing and retrieving data from the REU.

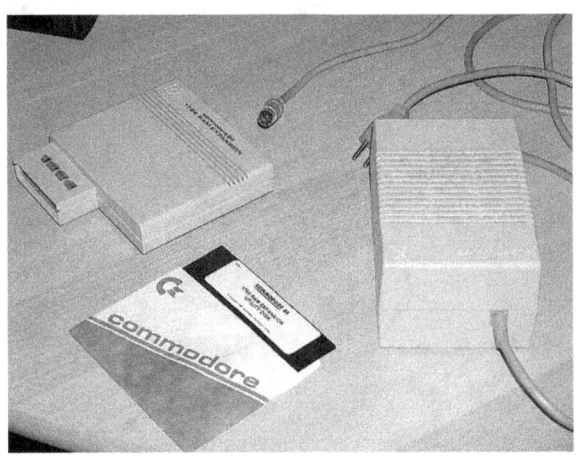

1764 REU with Utility Disk and 2.5 ampere power supply

Officially, only the 1700 and 1750 were supported on the C128. The 256 kB model, the 1764, was released for the C64 at the same time. However, aside from a bundled 2.5 ampere C64 power supply unit (the factory unit could not support the 1764), there were only minor differences between the three models.

In practice, the difference between the 1764 and the earlier units had little effect on compatibility, and people used 1700s and 1750s successfully with the C64, and 1764s successfully with the C128, although the C64's stock power supply was inadequate to reliably handle the power load of any of them. Some dealers unbundled the 1764 and the power supply in order to sell the power supply to C64 users, and/or upgrade the 1764 to 512 kB.

Because of memory chip shortages in the late 1980s, the 1750 was only produced in small quantities. However it was not difficult to upgrade a 1700 or 1764 to 512 kB. Several firms did this commercially, either selling upgraded units or upgrading customer-supplied units.

In the early 1990s, DIY modification schemes to increase the capacity of an REU to one megabyte or higher appeared on various online services.

2.22.2   Model differences

The 1700s circuit board was identical to that of the 1750, and a trace marked J1 indicated the size of the chips used. On the 1750 and 1764, this trace was cut.

The 1700 and 1750 had a resistor at position R4 that, according to Commodore engineer Fred Bowen, compensated for subtle differences in the expansion port on the C64 and C128. The 1764 lacked that resistor. Bowen and other CBM engineers recommended against using a 1764 with a C128 unless the resistor was added, or a 1700/1750 with a C64 unless the resistor was removed.

It was possible to check for the presence of a 1750 by reading memory address $DF00's bit 4, which was 1 on a 1750, and 0 on a 1700 or 1764. However, since this procedure would not distinguish between a 1700 and a 1764, many programmers wrote to the RAM itself to find out the amount of memory installed.

2.22.3   REU software support

Very little software made use of the REUs. Like other add-on products from Commodore, their relatively small installed base relative to the huge installed base of the C64 made software developers hesitant to invest much time and effort in supporting it, and the lack of commercial support kept sales lower than they otherwise might have been.

The REUs came with software to utilize the extra memory as a RAM disk, but the RAM disk's compatibility with commercial software was spotty, as some commercial software relied heavily on various quirks of the Commodore 1541 floppy drive. Additionally, many commercial programs simply overwrote the memory space occupied by the RAM disk software.

The GEOS operating system had built in support for the REU as a RAM disk, as did the C128's version of CP/M, and some disk copy programs used the REU to facilitate high-speed copying with a single disk drive. GEOS as well

as other programs even used the REU for quick memory transfers within the host machine's main memory by storing a memory block into the REU and then fetching it back to another location. Using this method, only the actual data to be transferred needed to travel on the machine's data bus— unlike the ordinary method, which had the computer's CPU do the transfer, thus spending at least three quarters of the bus capacity on instruction fetches and only one quarter or less on payload data.

Due to its high speed relative to Commodore's floppy drives or even the commercially available hard drives, the REU also became popular with BBS operators.

2.22.4   See also

- Super 1750 Clone
- geoRAM
- RAMLink
- Creative Micro Designs 2 MB REU, (**1750 XL**)

2.22.5   External links

- REU Games & Utilities
- REU Programming documentation
- Source code of RAMDOS, a RAM disk program for the Commodore RAM Expansion Unit

2.23   Commodore SX-64

Two Commodore SX-64 computers showing their SX-64 BASIC 2.0 startup screens. (Note the white screen background color.)

The **Commodore SX-64**, also known as the **Executive 64**, or VIP-64 in Europe, is a portable, briefcase/suitcase-size "luggable" version of the popular Commodore 64 home computer and holds the distinction of being the first full-color portable computer.[1]

The SX-64 features a built-in five-inch composite monitor and a built-in 1541 floppy drive. It weighs 10.5 kg (23lb). The machine is carried by its sturdy handle, which doubles as an adjustable stand. It was announced in January 1983 and released a year later, at US$ 995 (about $2,250 in 2014).[2][3]

2.23.1 Description

Aside from its built-in features and different form factor, there are several other differences between the SX-64 and the regular C64. The default screen color is changed to blue text on a white background for improved readability on the smaller screen. This can cause compatibility problems with programs that assume the C64's default blue background. The default device for load and save operations is changed to the floppy drive.

The Datasette (cassette) port and RF port were omitted from the SX-64. Because it has a built-in disk drive and monitor, Commodore did not perceive a need for a tape drive or television connector. However, the lack of a Datasette port poses a problem for a number of C64 Centronics parallel printer interfaces, since several popular designs "borrowed" their +5V power supply from the port. This was not an issue for later interfaces which were supplied with an AC adapter power supply, or those which can use the +5V line supplied by the Centronics port (Pin #18) on the printer itself, if the printer implements it. Alternatively, a +5V supply is also available from the joystick ports.

The audio/video port is still present, so an external monitor can still be used;[4] it and the built-in monitor display the same content.[5] Differences electrically and in placement on the board, means that there are compatibility problems with some C64 cartridges.

The original SX-64's (built in) power supply limits the machine's expandability.

Compatibility with Commodore RAM Expansion Units varies. Early SX-64 power supplies cannot handle the extra power consumption from the REU. The physical placement of the cartridge port can prevent the REU from seating properly. The 1700 and 1750, 128K and 512K units intended for the C128, are said to work more reliably with the SX-64 than the 1764 unit that was intended for the regular C64. Some SX-64 owners modify Commodore REUs to use an external power supply in order to get around the power supply issues.

A version of the SX-64 with dual floppy drives, known as the **DX-64**, was announced, but the press reported by early

1985 that plans for its release had been suspended.[4] A few have been reported to exist, but it is very rare. Instead of an extra floppy drive, a modem could also be built in above the first drive. Some hobbyists installed a second floppy drive themselves in the SX-64's empty drive slot. Later SX-64 units (from GA4 and on) use the larger power supply intended for the DX-64.

A version with a monochrome screen called the **SX-100** was announced but never released.

2.23.2 History

The SX-64 did not sell well, and its failure has been variously attributed to its small screen, high weight, bad marketing, and smaller business software library than that of its competitors, which included the Osborne 1 and Kaypro II (Zilog Z80 CPU, CP/M OS) and Compaq Portable (16-bit CPU, MS-DOS).

The exact number of SX-64 sold from 1984 to 1986, when it was discontinued, is unknown. The serial numbers of over 130 SX-64s from series GA1, GA2, GA4, GA5 and GA6, with serial numbers ranging over 49,000 for series GA1, 1,000 for GA2, 17,000 for GA4, 11,000 for GA5, and 7,000 for GA6 have been reported[6]

Some would-be buyers waited instead for the announced DX-64, which never became widely available due to the slow sales of the SX-64, creating a Catch 22 situation similar to that endured by Osborne after announcing an improved version of its computer. The SX-64 did however gain a following with user groups and software developers, who could quickly pack and unpack the machine to use for copying software or giving demonstrations.

2.23.3 Reception

Ahoy! favorably reviewed the SX-64, stating that the keyboard was better than the 64's, the monitor "isn't hard to read at all", and the disk drive was durable enough for travel. While criticizing the lack of any provision for internal or external battery power, the magazine concluded that the average $750-800 retail price was "worth every penny!".[4]

2.23.4 Technical information

Like the Commodore 64, except the following:

- Built-in storage: 170 kB 5¼" floppy disk drive (internal version of the Commodore 1541)

- Built-in display: 5" inch (127 mm) composite color monitor (CRT)

- Keyboard: Separate unit, connected by cord to CPU unit

- Cartridge port: Placed on top of CPU unit, w/spring-loaded fold-in lid, cartridges inserted vertically (vs horizontally into back of C64)

- I/O connectors:

  - Serial interface (rear)

  - Video out connector (rear)

  - User Port (rear)

  - Cartridge Port (beneath two spring-loaded flaps on the case top)

  - No Datassette interface

  - No RF modulator & connector

  - Non-standard 25-pin keyboard connector below right side of front panel. The connectors are similar but not identical to D-subminiature connectors and notoriously hard to find today

  - Standard three-prong IEC C14 AC power connector (vs C64 DIN plug to "power brick" PSU)

- Power supply: Internal unit with transformer and rectifiers (vs external C64 PSU)

- Extra features: Floppy disk storage compartment above disk drive which could be used to build in an extra floppy drive or compatible sized modem

2.23.5 References

[1] Commodore SX-64 Portable

[2] Commodore SX-64 portable computer

[3] Mace, Scott (February 6, 1984). "Commodore introduces new family of computers". *InfoWorld* (Menlo Park, CA: Popular Computing) **6** (6): pp 11–12. ISSN 0199-6649. "[Don Richards, Commodore USA president,] also said that the SX-64 computer, a $995 portable version of the Commodore 64 with built-in color monitor, has been a sellout everywhere."

[4] Benford, Tom (February 1985). "SX-64 Portable Computer". *Ahoy!*. pp. 37–38. Retrieved 15 October 2013.

[5] Crane, David (1985-01-21). "Ghostbusters demo". *The Computer Chronicles*. PBS.

[6] database at SX64.net

2.23.6 External links

- SX64 Dot Net

- SX-64 schematics (PDF format, zipped)

- Commodore SX-64 Paper Model

- C64 Preservation Project Preserving original C64 hardware and software

2.24 Compunet

Compunet was a United Kingdom based interactive service provider, catering primarily for the Commodore 64 but later for the Commodore Amiga and Atari ST. It was also known by its users as *CNet*.

It ran from 1984 before closing down in May 1993.

2.24.1 Overview

Compunet hosted a wide range of content, and users were permitted to create their own sections within which they could upload their own graphics, articles and software. A custom editor existed in which the "frames" that made up the pages could be created either offline or when connected to the service. The editor's cache allowed users to quickly download a set of pages, then disconnect from the service in order to read them, thus saving on telephone costs.

The user interface used a horizontally scrolling menu system, known as the "duck shoot", and navigation was essentially "select and click" with the ability to jump directly to pages with the use of keywords. Content could be voted upon by the users.

The service had many features which were considerably ahead of its time, especially when compared to the Internet of today:

- Pricing of content (Optional. Users could price their own content).

- Voting on content quality.

- "Upload anywhere" of content: programs, graphics and text (Unless a section was protected).

- Software could be dongle protected (the custom modem doubled as the dongle in this instance).

- WYSIWYG editing of content.

- Chat room (known as *Partyline*), which allowed users to create their own rooms (similar principles have been shown in IRC).

The server hosted Multi-User Dungeon (MUD) (by Richard Bartle), Federation II, and Realm. The first two of these games continue to run on the Internet today.

Games creator Jeff Minter and musician Rob Hubbard, along with various members of the demo scene, had a presence on the network.

2.24.2 History

In 1982, Commodore UK decided to construct a nationwide computer network for the use of teachers. The Commodore PET computer had been very successful. Nick Green developed the specification of what became *PETNET* with David Parkinson and Mike Bolley of *Ariadne Software* in The Albany pub (see "PETNET - data transmission system" in "Microcomputers in education" ed Dr I.C.H. Smith 1982 John Wiley ISBN 0-85312-424-8).

In the Summer of 1982 Keith Hall of Commodore secured the money to commission the prototype which was run on an ADP DEC-10 machine. *Ariadne Software* wrote the software in 6502 Assembler for the client and FORTRAN for the host. The X25 packet protocol was modified to provide error correction for all file transfers. At ADP's suggestion reliable uploading was achieved by using temporary file names which were changed to user file names when the last byte had been correctly received by the host. Nick Green sought partners who could provide local call access and Host facilities. Alan Carmichael, Graham Craigie and Robert Foot of ADP joined the project.

Around this time the BBC Micro was released and gained enormous popularity within the UK education system. Commodore's 64 was seen as the "more bangs per buck" American alternative. PETNET became Compunet which was aimed to support consumer and educational users.

Nick Green specified a secure modem based on the Viewdata chip set and the assembly language client software was ported into the modem and bundled with the Commodore 1541 disk drive. After the first year Commodore was bought out and Compunet Teleservices Ltd became an independent company. Compunet culture was covered in the first issue of Commodore Disk User, which shipped software on its cover disk.

ADP provided the initial DEC-10 mainframe, as well as the local-access dial-up points. But this was very expensive and a scheduled migration to a VME bus based multi-micro machine was successfully undertaken. New local dial-up points were provided by ISTEL (on their *Fastrak* network). After a management buy out *ISTEL* was sold to AT&T. This led to the failure of *ISTEL* technical support and an upgrade of local access to 2400/2400 baud. The best efforts of Ariadne and Compunet staffers Jason Gold and Mark Clarke came

to nothing as the English legal system failed to protect Compunet's contracts. This meant a higher cost nationwide rate call for most users. A third move of the Compunet Host to Camden in North London was undertaken with Nick Green now board chair and MD.

By this time client software was ported to the Commodore Amiga and Atari ST and a teletype compatible version of the service using BBS scrolling text was introduced aimed at integration with the Internet and PCs.

Compunet ceased trading in May 1993, when the company went into receivership for non-payment of VAT after the sudden short illness and death of Jim Chalmers, their sole practitioner accountant. He was negotiating a VAT refund at the time. Immediate barrister intervention failed despite £250,000 of debt asset.

2.24.3 Subscription model

Compunet charged a quarterly subscription, and telephone call costs were in addition to this. Typical off-peak charges would be £0.80UKP per hour.

Premium services incurred additional charges, which required the user to first place money in their account. These services included:

- Private e-mail (some free quota was provided).

- Uploading content.

- Custom banners.

- Customised user name (instead of, for example, **'abc3'**).

- Access to chat and gaming services.

2.24.4 Technology

Client

For the Commodore 64, Compunet provided a custom 1200/75 baud modem (affectionately known as the "brick") which utilised the machine's cartridge port. As well as the usual modem features, the device had a custom ROM which contained the rudiments of the software required to access the service. This software could be updated automatically upon connection to the service.

Out of the box, the modem was unable to connect to standard Bulletin board systems unless an optional software package was purchased.

The modem was programmed with a unique ID. This allowed it to work as a dongle to help prevent piracy of protected software.

The custom nature of the technology hindered Compunet to a degree. The graphical design was very much keyed into the Commodore 64's graphical capabilities. Although this was more powerful than the Viewdata systems such as Prestel, it meant porting was difficult. However, software was later made available for the Amiga (1987) and Atari ST (1988). A PC version was developed in-house but never made publicly available.

The Amiga and Atari ST versions both emulated the graphics and interface of the original Commodore 64. However, the PC version was teletype in nature, utilising Kermit for file transfers.

Server

The host server was a DEC-10 at launch, which ran Compunet as a time-slice. ADP provided the mainframe, as well as the local dial-up points, which allowed users all over the country access for the cost of a local telephone call.

Specification:

- 1 megaword 36 bit RAM (upgraded for Compunet).

- £50,000 per month running costs (including the local-rate telephone call facilities).

- Compunet host software written in Fortran, by *Ariadne Software* and further developed by Robert Foot with chat and real-time user and management accounting.

When ADP announced it was to shut down its DEC-10 network in Great Portland Street, Central London, Nick Green then consultant to Compunet and Mark Clarke (ex Commodore guru) researched closely coupled multi-micro architectures. Compunet was rewritten in C and migrated to a VME rack configured by Cambridge Micro Computers in the Park Royal Industrial Estate. It ran under OS-9 with a single 25 MHz master board.

Specification:

- VME bus.

- 1 x 6820 for disk access.

- 10 megabytes RAM.

- 4 x 200 megabyte hard disks for storage.

- 3 x 6810s (5 megabytes RAM each) for communications.

- 52 simultaneous connections.

2.24.5 Sources

- Article: "CNET - Moving with the times"

- Direct discussion with Nick Green (ex-Chairman of Compunet).

2.24.6 External links

- Commodore 64 Apocalypse - Compunet pages, photos & interviews

- Federation II

- MUD2

- Gnome's Computers (an article that originally appeared on Compunet)

2.24.7 See also

- Micronet 800

2.25 Creative Micro Designs

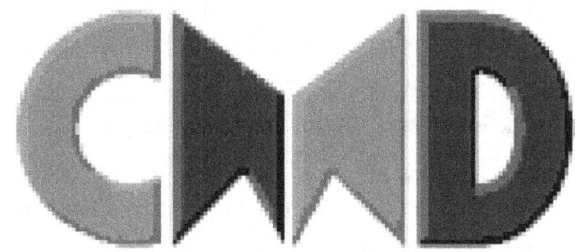

the CMD logo

Creative Micro Designs (CMD) was founded in 1987 by Doug Cotton and Mark Fellows. It is a computer technology company which originally developed and sold products for the Commodore 64 and C128 8-bit personal computers. After 2001 it sold PCs and related equipment.

2.25.1 History

CMD's first product, JiffyDOS, was developed from 1985 onwards by Mark Fellows. An updated disk operating system, it maintained broad compatibility with Commodore floppy drives' DOS while offering much increased read write access.

CMD stopped selling Commodore products in 2001. In July of that year, programmer Maurice Randal was sold an exclusive license to produce and sell the Commodore related products. His company Click Here Software Co supplied the products until around 2009.[1]

In 2010, Jim Brain acquired the license to supply JiffyDOS. Since January of that year, he has sold the product via his web shop Retro Innovations.

2.25.2 Products

- SuperCPU - A 65816 CPU 8/16-bit upgrade for the C64 and C128 released on May 4, 1997, with version 2, the C128 compatible version, being launched in 1998.[2]

- RAMLink - A 'fast' solid-state RAM-Disk that would plug into the cartridge port of the C64 or C128 which added between 1 Megabyte and 16 Megabytes. The C64 version typically required a 'timer jump clip'. The RAMPort allowed it to work with the Commodore 17xx RAM Expansion Units

- FD series - The FD2000 used 'High Density' Disks of up to 1.6 Megabytes of storage, with the FD4000 using 'Enhanced Density' Disks of up to 3.2 Megabytes of storage

- HD series - SCSI Hard drives of between 20 Megabytes and 4.4 Gigabytes using CMD's native partitioning system of 16 Megabytes per partition

- JiffyDOS - Adds DOS Wedge commands for easier functionality via BASIC command prompt

- Swiftlink/Turbo232 - Adds dial-up modems to your Commodore 64 or 128 of up to 33.3kbit/s (Swiftlink) or 56.6kbit/s (Turbo232)

- 1750 XL - a Commodore 17xx REU clone in two flavours adding either 512 Kilobytes or 2 Megabytes

- SuperRAMCard - Works in conjunction with the SuperCPU to add between 1 Megabyte and 16 Megabytes of directly accessible memory using the 65816 processor

- SmartTRACK/SmartMOUSE - An 'intelligent' Commodore 1351 3-buttoned mouse or trackball which had 2K of RAM and a battery-backed Y2K compliant Real Time Clock which was GEOS compatible

2.25.3 References

[1] Old Computers Museum

[2] C64 CPU Speed up Cartridges - The History

2.25.4 External links

- The Unofficial CMD Homepage

- JiffyDOS at Brain Innovations officially licensed JiffyDOS products

2.26 DolphinDOS

DolphinDOS is a hardware expansion for the Commodore computers and floppy disk drives like Commodore 1541, 1541-II, Commodore 1571. It combined the disk controller side RAM expansion and firmware replacement with computer side KERNAL replacement and additional Parallel connection between the disk drive controller and the computer.

2.26.1 See also

- Commodore DOS

- Commodore 64

- Commodore 128

- Commodore International

2.27 Epyx Fast Load

The **Epyx Fast Load** is a floppy disk fast loader cartridge made by American software company Epyx in 1984 for the Commodore 64 home computer. It was programmed by Epyx employee Scott Nelson, who later designed the Epyx Vorpal fastloading system for the company's games.

2.27.1 Description

Epyx Fast Load allows programs to load from the Commodore 1541 disk drive approximately five times faster than the normal speed. Since it is stored on a cartridge, and thus provides instant access without requiring any hardware modification of the C64 or disk drive, the Fast Load quickly became a very popular peripheral among C64 users.

In addition to disk acceleration, the cartridge also provides a built-in version of the Commodore DOS Wedge. This dramatically reduces the number of keystrokes needed to load or save files or perform disk operations, and makes the cartridge even more convenient.

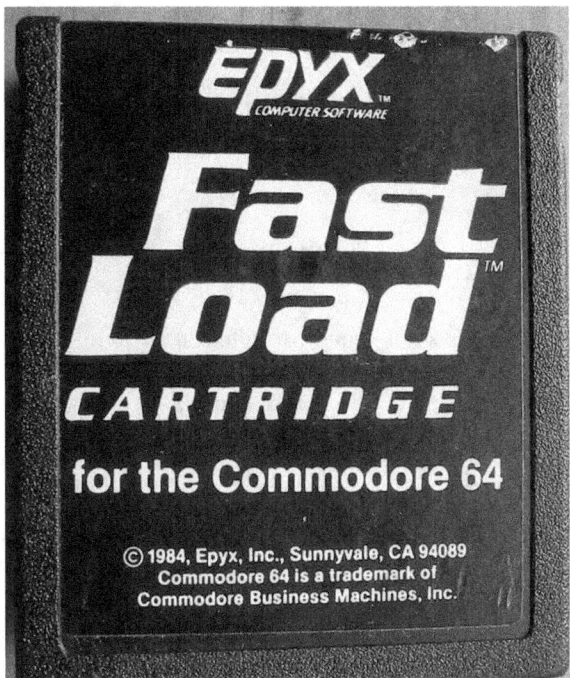

*The **Epyx Fast Load** cartridge was a bestseller for the C64.*

Epyx Fast Load incorporates a machine language monitor. Although it does not include an assembler, as most "standard" C64 ML monitors do, it includes a wide array of powerful debugging tools. These include disassembly, single-stepping, and an automatic machine code relocator.

A crude disk editor is also included with the cartridge, which displays raw data from floppy disks in classical hex+ASCII split screen mode. Among other things, the disk editor can be used to enter cheat codes and do the home computer variant of ROM hacking.

In the unusual case of software that doesn't work with the Fast Load, the cartridge can be disabled via a menu command, thus avoiding the need to physically remove and reinsert it.

2.27.2 Reception

In a review of three Commodore 64 fast loaders, *Ahoy!* wrote that the product "is surprisingly transparent to all the forms of commercial copy protection we have looked at ... In terms of greatest convenience and speed, we place our bets on the Epyx Fast Load".[1]

2.27.3 References

[1] Kevelson, Morton (May 1985). "Disk Spinners, Part II". *Ahoy!*. pp. 33–38. Retrieved 16 October 2013.

2.27.4 External links

- Epyx Fast Load instruction manual from Project64

- RUN Magazine Issue 56

2.28 geoRAM

geoRAM from Berkeley Softworks was a memory expansion peripheral for use on the Commodore 64 computer with GEOS operating system. geoRAM was created by Dave Durran.[1][2]

During the chip shortages of the 1980s, Commodore could not produce enough of its RAM Expansion Units (they eventually cancelled them). The GEOS operating system relied heavily on extra RAM and so the company behind GEOS produced their own memory expansion cartridge, called the **geoRAM**.[3][4]

Using a mapped-in page scheme, the geoRAM was much slower in operation than the DMA-driven REU cartridges; hence, not much software other than GEOS actually supported it.

2.28.1 References

[1] The GEOS Column: Closeup on GEORAM, By Robert Bixby, Page 14, Issue 83, 1990 May, Compute Gazette

[2] *Dave Durran, Vice President, Hardware Development, Mr. Durran was a cofounder of Geoworks and served as Hardware Architect there from 1983 to 1998.*

[3] First look at: geoRAM

[4] 13.3.2. What is a geoRAM Unit?, *When Commodore REUs became hard to find several years back, Berkeley Softworks introduced geoRAM, which is a 512K RAM expander. This RAM expander gives you all of the advantages of a 1750 with GEOS. However, it is not 1750 compatible, so it will not work like a 1750 outside of GEOS; it is transparent to other programs. (As a caveat on this, see the info on RAMLink). DesTerm128 2.0 will not work with a geoRAM plugged in. A special version of GEOS 2.0 (which is bundled with geoRAM) is necessary to use geoRAM. No additional power supply is necessary to use geoRAM.*, COMP.SYS.CBM: General FAQ, v3.1 Part 8/9

2.29 Human Engineered Software

This article is about an American video game publisher. For an Australian video game distributor, see Home Entertainment Suppliers.

Human Engineered Software (**HES**, also known as **HesWare**) was an American home computer software and hardware developer/publisher during the 1980s, who concentrated on the Commodore 64 and the Atari 8-bit.

2.29.1 History

The company was located in Brisbane, California. Published titles included numerous games as well as educational and productivity programs. Among them were *Project Space Station*, *Mr. TNT*, Turtle Graphics by David Malmberg, several Jeff Minter games (Llamasoft), such as *Attack of the Mutant Camels*, *Gridrunner*, *Hes Games*, and HesMon, Graphics BASIC, 64Forth (a cartridge-based Forth implementation), and the HesModem and HesModem II. At one point, HES was the largest single-source supplier of software for the Commodore 64.

The company was started by Jay Balakrishnan and Cy Shuster in 1980, in Jay's apartment in Los Angeles. Typical of his creativity, Jay took down the door to his bedroom, put it across two file cabinets, and used that as a desk for his development (winding the cables around the doorknob!). With incredible amounts of research into the PET ROM, Jay wrote the first 8K 6502 Assembler, HESbal (HES Basic Assembler Language) in BASIC, and an accompanying text editor, HESedit. Having HESbal allowed numerous creative follow-on products, such as HEScom, software and a user port cable that allowed VIC20 programs to be saved to a PET hard disk (since the first VIC20 didn't have a hard disk). Cy soldered the HEScom cables in his garage and wrote HESlister, a print utility for BASIC programs, that he ported from a TRS-80 Model I to the PET, to the VIC, and later to the IBM PC. HESware published OMNIWRITER, a word processor for the Commodore 64.

Game writers Lawrence Holland and Ron Gilbert, later to be famous for their work at LucasArts, started their careers at HES.

By early 1984 *InfoWorld* estimated that HES was tied with Broderbund as the world's tenth-largest microcomputer-software company and largest entertainment-software company, with $13 million in 1983 sales.[1] In October 1984, however, HES was acquired by Avant Guard Publishing Corp. and thus, already in deep financial troubles, avoided having to declare bankruptcy.

2.29.2 References

[1] Caruso, Denise (1984-04-02). "Company Strategies Boomerang". *InfoWorld*. pp. 80–83. Retrieved 10 February 2015.

2.30 IDE64

The **IDE64** interface cartridge is an expansion port device for connecting ATA(PI) devices to the C64 or C128 computers.

2.30.1 Hardware

There were several somewhat different versions[1] of this cartridge over the years. The interface was designed by Tomas Pribyl and Jan Vorlicek in 1994. Today Josef Soucek is working on the design.

IDE64 v1.1 In 1997 the first public version of the cartridge appeared. The logic was fitted into 2 ispLSIs, the operating system was burned into a 32 kB EPROM, and there was 16 kB of RAM storage for buffers. There was a DS1302 real-time clock included which could keep the settings and hold the time backed by a battery. Beside the parallel ATA connector also an expansion port pass through was included to allow attaching of other cartridges.

IDE64 v2.1 Around 1999 the first "modern" version of the cartridge came out, merging the logic into one big ispLSI PLD. Also the EPROM was upgraded to a 64 kB EEPROM to allow operating system upgrades without special equipment. The RAM was extended to hold 28 kB of buffers and internal variables. There was a new connector on board, called ShortBus. It was meant for hardware expansions like LCD displays,[2] 7-segment display,[3] but later also more sophisticated expansions appeared. This was also the first version to include SuperCPU compatibility.

IDE64 v3.1 In 2001 a redesigned version of the 2.1 cartridge was made, this version drops the expansion port pass-through.

IDE64 v3.4 CompactFlash cards became widely popular, so the cartridge was redesigned in 2004 to include a CF socket. This allowed a really mobile, and small storage possibility for the C64, without additional devices and power supplies.

IDE64 v3.4+ In 2005, with a small hardware change, the EEPROM was upgraded to 128 kB, which allowed to hold two versions of the operating system for both a standard C64 and SuperCPU. This was selectable with a small switch, and solved the reflashing problem for those with SuperCPU equipped systems.

IDE64 v4.1 The pre-release happened in 2008 August,[4] but it was not available until 2009 March. The hardware was redesigned to use surface-mounted parts and a more up-to-date ispMACH CPLD, which resulted in a much shorter board. An USB serial fifo chip was added for fast PCLink connections, and an Amiga clock-port for connecting additional devices. The card slot on this version is separated from the parallel ATA port, which is more compatible to strange CF cards. The I/O interface was changed to support 128 kB operating system and was tweaked for slightly faster data transfer speeds.

2.30.2 ShortBus expansions

ETH64 A LAN91C96 chip based Ethernet card. It is supported by Contiki, Wings and maybe some other software. It can also be used for PCLink connection.

DUART This is a XR68C681 based dual RS-232 card featuring, mostly used for PCLink connection. It is supported by Contiki, Wings, Novaterm 9.6 and maybe some other software.

DigiMAX It is a MAX506 based 4 channel 8-bit digital-to-analog converter card, can be used as "sound card", as the output comes out on two jack plugs. This card is supported by Modplay, Wings and maybe some other programs.

ETFE This is a CS8900 based Ethernet card, just like the popular RR-Net, but the v1.1 version works only in TFE compatible mode. It can be used for PCLink, has a Contiki driver, and work with software designed for the original TFE card. For the next version (v1.2) a jumper is promised to simulate a RR-Net card.

2.30.3 External links

- The webpage of the authors of the cartridge

- The webpage of the operating system used

- The news webpage of the IDE64 project

- IDE64 on C64 Wiki (German)

- IDE64 on HupWiki (Hungarian)

2.30.4 References

[1] Pictures of IDE64 versions

[2] LCD display driver source

[3] Hexcard for IDE64

[4] Árok Party 2008 szerzemények #2 - avagy Agyament Hardverek #1 (Hungarian)

2.31 Individual Computers

Individual Computers is a German computer hardware company specializing in retrocomputing accessories for the Commodore 64, Amiga, and PC platforms. Individual Computers produced the C-One reconfigurable computer in 2003. The company is owned and run by Jens Schönfeld.

2.31.1 Products

- Catweasel – Universal format floppy disk drive controller card

- Retro Replay – Improved version of the C64 Action Replay cartridge

- Clone-A – Amiga in FPGA website (coming soon?)

  - See the PDF extract of Total Amiga Magazine issue 25

- MMC64 – MMC and SD Card reader cartridge

- MMC Replay – MMC64 and Retro Replay combined in one cartridge, with some improvements

- Micromys – An adapter that allows connecting PS/2 compatible mice (including wheel-support) to C64 and Amiga joystick-ports (and all other computers that share the same pin-configuration).

- Amiga clock port compatible addons for MMC64, Retro Replay and MMC Replay:

  - RR-Net: A C64-compatible Network-Interface. Comes in 2 shapes, the old long RR-Net fits Retro Replay and MMC64 (though partly blocking the latter's passthrough expansionport), the new L-shaped RR-Net2 fits MMC64 and MMC Replay and was built with MMC Replay in mind.

  - Silver Surfer: Highspeed RS232 Interface for the C64. Fits onto Retro Replay, MMC64/Replay compatibility unknown.

  - mp3@c64: hardware mp3 decoding from SD card. Made for MMC64, Retro Replay and MMC Replay compatibility unknown.

- Keyrah – An interface that allows the connection of Commodore keyboards to USB-capable computers

- C-One – reconfigurable computer

- X-Surf – network card

- C64 Reloaded - A 1:1 rebuilt C64 motherboard with less power consumption

2.31.2 External links

- Individual Computers official website

- Individual Computers Product Information Wiki

2.32 Indus GT

The **Indus GT** is a floppy disk drive that was made by Indus Systems of California, USA during the early 1980s for Apple II series and Atari home computer and later for the Commodore platforms of the day. It came in a black casing with a smoke plexiglas cover over the drive mechanism. Behind the cover is an 8-segment LED track and sector display. It was considered as a high-quality unit, regarded by many as the best floppy drive available for 8-bit Ataris. It was advertised as being over 400% as fast as the Commodore 1541 (which is a false claim, as it is no faster) and has an internal "ROM drive" with DOS utility software, but suffers, as do all third party Commodore drives, from being less than 100% 1541-compatible.

2.32.1 External links

- RUN Magazine Issue 20

2.33 ISEPIC

The **ISEPIC** from Starpoint Software in USA is an extension cartridge which was introduced in June 1985 for the Commodore 64. It offers the capability to memory dump software regardless of the implementation scheme or storage medium. The resulting snapshot can be tested before saving.

Snapshots require ISEPIC to run.[1] The cartridge vanished at the end of 1985 but still sold 20,000 units around the world, mostly by word of mouth at local computer clubs and niche magazines.

A major factor is the 2 kB RAM that could be reprogrammed And thus allowed the user to change its functionality. The 2 kB RAM is memory banked into a 256-byte page at 0xDF00 – 0xDFFF.

2.33.1 Reception

Ahoy! in October 1985 stated that "the ramifications of [ISEPIC] are startling, to say the least". While warning readers against violating copyright, the magazine discussed the cartridge's ability to both produce snapshots that required the cartridge to boot, and help users modify snapshots to produce standalone versions of programs ("In the tradition of the true hacker, these routines also display the *Isepic* logo while booting the program").[2] In a review of two other memory dumpers in March 1986, the magazine stated that they were superior: "While ISEPIC did not do bad as a forerunner, it does not measure up to the products reviewed here" which, among other improvements, produced standalone snapshots.[1]

2.33.2 References

[1] Kevelson, Morton (1986-03). "Memory Dumpers for the C-64". *Ahoy!*. pp. 60–66. Retrieved 23 July 2015. Check date values in: |date= (help)

[2] Kevelson, Morton (1985-10). "Isepic". *Ahoy!*. pp. 71–73. Retrieved 27 June 2014. Check date values in: |date= (help)

2.33.3 See also

- Trilogic Expert Cartridge - A later cartridge with 8 kB RAM

2.33.4 References

2.33.5 External links

- c64.org - ISEPIC

2.34 KERNAL

This article is about Commodore's 8-bit OS software. It is not to be confused with Kernel (disambiguation).

The **KERNAL**[1] is Commodore's name for the ROM-resident operating system core in its 8-bit home computers; from the original PET of 1977, followed by the extended but strongly related versions used in its successors: the VIC-20, Commodore 64, Plus/4, C16, and C128.

2.34.1 Description

The Commodore 8-bit machines' KERNAL consists of the low-level, close-to-the-hardware OS routines roughly equivalent to the BIOS in IBM PC compatibles (in contrast to the BASIC interpreter routines, also located in ROM) as well as higher-level, device-independent I/O functionality, and is user-callable via a jump table whose central (oldest) part, for reasons of backwards compatibility,[2] remains largely identical throughout the whole 8-bit series. The KERNAL ROM occupies the last 8 KB of the 8-bit CPU's 64 KB address space ($E000-$FFFF).

The jump table can be modified to point to user-written routines, for example rewriting the screen display routines to display animated graphics or copying the character set into RAM. This use of a jump table was new to small computers at the time.[3]

The Adventure International games published for the VIC-20 on cartridge are an example of software that uses the KERNAL. Because they only use the jump table, the games can be memory dumped to disk, loaded into a 64, and run without modification.[4]

The KERNAL was initially written for the Commodore PET by John Feagans, who introduced the idea of separating the BASIC routines from the operating system. It was further developed by several people, notably Robert Russell, who added many of the features for the VIC-20 and the C64.

2.34.2   Example

A simple, yet characteristic, example of using the KERNAL is given by the following 6502 assembly language subroutine[5] (written in ca65 assembler format/syntax):

CHROUT = $ffd2 ; CHROUT sends a character to the current output device CR = $0d ; PETSCII code for Carriage Return ; hello: ldx #0 ; start with character 0 next: lda message,x ; read character X from message beq done ; we're done when we read a zero byte jsr CHROUT ; call CHROUT to output char to current output device (defaults to screen) inx ; next character bne next ; loop back while index is not zero (max string length 255 bytes) done: rts ; return from subroutine ; message: .byte "Hello, world!" .byte CR, 0 ; Carriage Return and zero marking end of string

This code stub employs the CHROUT routine, whose address is found at address $FFD2 (65490), to send a text string to the default output device (e.g., the display screen).

2.34.3   The name

The KERNAL was known as *kernel*[6] inside of Commodore since the PET days, but in 1980 Robert Russell misspelled the word in his notebooks forming the (non-)"word" *kernal*. When Commodore technical writers Neil Harris and Andy Finkel collected Russell's notes and used them as the basis for the VIC-20 programmer's manual, the misspelling followed them along and stuck.[7]

According to early Commodore myth, and reported by writer/programmer Jim Butterfield among others, the "word" KERNAL is an acronym (or maybe more likely, a backronym) standing for *Keyboard Entry Read, Network, And Link*, which in fact makes good sense considering its role. Berkeley Softworks later used it when naming the core routines of its GUI OS for 8-bit home computers: the GEOS KERNAL.

2.34.4   On device-independent I/O

Surprisingly, the KERNAL implemented a device-independent I/O API not entirely dissimilar from that of Unix or Plan-9, which nobody actually exploited, as far as is publicly known. Whereas one could reasonably argue that "everything is a file" in these latter systems, you could easily claim that "everything is a GPIB-device" in the former.

Due to limitations with the 6502 architecture at the time, opening an I/O channel requires three system calls. The first typically sets the logical filename through the SETNAM system call. The second call, SETLFS, establishes the GPIB/IEEE-488 "device" address to communicate with. Finally OPEN is called to perform the actual transaction. The application then used CHKIN and CHKOUT system calls to set the application's current input and output channels, respectively. Applications may have any number of concurrently open files (up to some system-dependent limit; e.g., the C64 allows for ten files to be opened at once). Thereafter, CHRIN and CHROUT prove useful for actually conducting input and output, respectively. CLOSE then closes a channel.

Observe that no system call exists to "create" an I/O channel, for devices cannot be created or destroyed dynamically under normal circumstances. Likewise, no means exists for seeking, nor for performing "I/O control" functions such as you'd find with ioctl() in Unix. Indeed, KERNAL proves much closer to the Plan-9 philosophy here, where an application would open a special "command" channel to the indicated device to conduct such "meta" or "out-of-band" transactions. For example, to delete ("scratch") a file from a disk, you typically will "open" the resource called "S0:THE-FILE-TO-RMV" on device 8 or 9, channel 15. Per established convention in the Commodore 8-bit world, channel 15 represents the "command channel" for peripherals, relying on message-passing techniques to communicate both commands and results, including exceptional cases. For example, in Commodore BASIC, you might find software not unlike the following:

70 ... 80 REM ROTATE LOGS CURRENTLY OPENED ON LOGICAL CHANNEL #1.   90 CLOSE #1 100 OPEN 15,8,15,"R0:ERROR.1=0:ERROR.0" 110 INPUT #15,A,B$,C,D 120 CLOSE #15 130 IF A=0 THEN

GOTO 200 140 PRINT "ERROR RENAMING LOG FILE:" 150 PRINT " CODE: "+A 160 PRINT " MSG : "+B$ 170 END 200 REM CONTINUE PROCESSING HERE, CREATING NEW LOG FILE AS WE GO... 210 OPEN 1,8,1,"0:ERROR.0,S,W" 220 ...

Device numbers, per established documentation, are restricted to the range [0,16]. However, this limitation came from the specific adaptation of the IEEE-488 protocol and, in effect, applies only to external peripherals. With all relevant KERNAL system calls vectored, programmers can intercept system calls to implement virtual devices with any address in the range of [32,256]. Conceivably, one can load a device driver binary into memory, patch the KERNAL I/O vectors, and from that moment forward, a new (virtual) device could be addressed. So far, this capability has never been publicly known as utilized, presumably for two reasons: (1) The KERNAL provides no means for dynamically allocating device IDs, and (2) the KERNAL provides no means for loading a relocatable binary image. Thus, the burden of collisions both in I/O space and in memory space falls upon the user, while platform compatibility across a wide range of machines falls upon the software author. Nonetheless, support software for these functions could easily be implemented if desired.

Logical filename formats tends to depend upon the specific device addressed. The most common device used, of course, is the floppy disk system, which uses a format similar to "MD:NAME,ATTRS", where M is a flag of sorts ($ for directory listing, @ for indicating a desire to overwrite a file if it already exists, unused otherwise.), D is the (optional) physical disk unit number (0: or 1: for dual-drive systems, just 0: for single-disk units like the 1541, et al, which defaults to 0: if left unspecified), NAME is a resource name up to 16 characters in length (most characters allowed except for certain special characters), and ATTRS is an optional comma-separated list of attributes or flags. For example, if you want to overwrite a program file called PRGFILE, you might see a filename like "@0:PRGFILE,P" used in conjunction with device 8 or 9. Meanwhile, a filename for the RS-232 driver (device 2) consists simply of four characters, encoded in binary format.[8]

Other devices, such as the keyboard (device 0), cassette (device 1), the display interface (device 3), and printer (device 4 and 5), require no filenames to function, either assuming reasonable defaults or simply not needing them at all.

2.34.5 Notes

[1] *Commodore 64 Programmer's Reference Guide*. Commodore Business Machines, Inc., 1982, p. 268

[2] The KERNAL jump table, used to access all the subroutines in the KERNAL, is an array of JMP (jump) instructions leading to the actual subroutines. This feature ensures compatibility with user-written software in the event that code within the KERNAL ROM needs to be relocated in a later revision.

[3] "Exploring the VIC-20".

[4] Kevelson, Morton (January 1986). "Speech Synthesizers for the Commodore Computers / Part II". *Ahoy!*. p. 32. Retrieved 17 July 2014.

[5] Many of the KERNAL subroutines (e.g., OPEN and CLOSE) were vectored through page three in RAM, allowing a programmer to intercept the associated KERNAL calls and add to or replace the original functions.

[6] The kernel is the most fundamental part of a program, typically an operating system, that resides in memory at all times and provides the basic services. It is the part of the operating system that is closest to the machine and may activate the hardware directly or interface to another software layer that drives the hardware

[7] *On The Edge: The Spectacular Rise and Fall of Commodore*, page 202.

[8] *Commodore 128 Programmers Reference Guide*, Commodore Business Machines, Inc., 1986, p. 382

2.35 Lt. Kernal

Lt. Kernal is a SCSI hard drive subsystem developed for the Commodore 64 and Commodore 128 home computers. The original design of both the technically complicated hardware interface and equally complex disk operating system came from Lloyd Sponenburgh and Roy Southwick of Fiscal Information, Inc., a now-defunct Florida-based turnkey vendor of minicomputer-based medical information systems.

Fiscal demonstrated a working prototype in 1984 and starting advertising the system for sale early in 1985. It immediately found a niche with some Commodore software developers and bulletin board SysOps due to its excellent performance and capacious storage (originally 10 megabytes and later extended to as much as 330 megabytes). The subsequent development of a multiplexing accessory allows one Lt. Kernal to be shared by as many as 16 computers, using a round robin scheduling algorithm. This made the use of the Lt. Kernal with multiple line BBSs practical. Later, streaming tape support, using QIC-02 tape cartridges, was added to provide a practical (though costly) backup strategy.

A key feature of the Lt. Kernal is its sophisticated disk operating system, which behaves much like that of the Point 4 minicomputers that Fiscal was reselling in the 1980s. A high degree of control over the Lt. Kernal is possible with simple typed commands, many of which had never been seen before in the 8-bit Commodore environment. This, along with a powerful keyed random access filing system, makes the Lt. Kernal perform at a level that was generally unmatched by any other hard drive system available for 8-bit Commodore computers.

Fiscal built the units to order until late 1986, at which time the decision was made to turn over the production, marketing and customer support to Xetec Inc. Fiscal continued to provide secondary technical support, as well as free DOS upgrades, until December 1991, at which time production of new Lt. Kernal systems ceased. Following the shutdown of Xetec in 1995, private support of the Lt. Kernal was carried on for several years by Ron Fick until his untimely death in 1999.

2.35.1   External links

- Lt. Kernal Data Archive

2.36   Magnum Light Phaser

The Magnum Light Phaser

The **Magnum Light Phaser** is a light gun created in 1987 for the ZX Spectrum computer. A version was also released for the Commodore 64/128. It was Amstrad's last peripheral for the video game console. The Magnum Light Phaser in many ways resembles the Light Phaser, the Sega Master System light gun, released in 1986. It was a Sinclair-branded Far Eastern product which was included in promotional bundles such as the "James Bond 007 Action Pack", along with a small number of lightgun-compatible games.

It was also available separately in a £29.95 pack along with six games. Only a few games bothered with lightgun compatibility (Operation Wolf, the original arcade gun game,

was the most notable) and fewer still were produced specifically for use with the Magnum. Even so, the lightgun was widely available, largely because Amstrad's bundling policy ensured wide distribution.

Software Creations created five exclusive games for the Commodore 64 package.

2.36.1   Supported Games

Spectrum:

- Bullseye
- James Bond 007
- Missile: Ground Zero
- Operation Wolf
- Robot Attack
- Rookie
- Solar Invasion

Bundled with the Commodore 64 version:

- Baby Blues
- Cosmic Storm
- Ghost Town
- Goosebusters
- Gunslinger
- Operation Wolf (replaces the NEOS mouse control option)

Bundled with the Commodore 64 Lightgun package, and compatible with the Magnum:

- Army Days
- Gangster
- Time Traveller
- Blaze-Out (compilation of Ocean game sequences with lightgun controls)

2.36.2   External links

- Crash review of the Magnum Light Gun and Games
- Light Phaser

2.37 MMC64

The **MMC64** is a cartridge for the C64 home computer, which plugs into the expansion port. It was developed in 2005 by Oliver Achten, production and sale is done by the Commodore hardware accessory company Individual Computers, although the MMC64 is sold by other retailers as well.

2.37.1 Hardware Features

The MMC64 serves as a read/write interface for FAT16 and FAT32 (since BIOS V1.04) formatted MMC and SD flash memory media. It can handle almost all media up to 4 GByte, however some cards are completely incompatible and result red or black screen in power up. The MMC64 is a solution for Commodore users to transfer data between the Commodore 64 and a PC. The PC does not need to have a parallel port, which is a requirement for the popular x1541 interface.

It has a pass-through expansion port, allowing to use another cartridge together with the MMC64.

The MMC64 also provides an Amiga style clock port for extra hardware, supporting the RRnet Ethernet interface and the "mp3@c64", which allows playing of mp3 files from the flash memory.

The MMC64 is sold "bare" without a case, which means that users who desire protection need to manually modify full cases to match its dimensions (and the dimension of pass-through expansion port Cartridges).

2.37.2 Software Features

Built-in features

The MMC64 has a flashable BIOS which is updated for new features, better compatibility and other improvements. There is also an alternative BIOS (software download link) available from another developer. Documentation and source code are provided to facilitate development of additional plugins and alternative BIOS images. Some later BIOS versions require to be renamed as mmc64v1b.upg to be read from system64 directory when pressing F5-key.

The cartridge has a file browser that is launched automatically upon start, although the auto-launch feature allows for different file browsers. It can execute .PRG files on the flash media right from the file browser and write .d64 disk image files to disk, although this is very slow with the built-in diskwriter, so most people use a faster plugin (software download link), although this plugin does not work on metal C128Ds due to timing issues. (The latest version of the BIOS, 1.03, removes the slow built-in .d64 writer in order to make space for more useful BIOS features; it is recommended that people use the much faster plugin for this task.)

There is a SID music file player built in that allows playing SID files from the HVSC on a real 6581 or 8580 SID chip (in the computer).

If the "R" key is pressed upon startup, the MMC64 reads the disk in the drive and saves it to the flash media as a disk image. This is slow compared to other solutions like the RR-Net based Warpcopy or the x1541 parallel port interface, but faster implementations are being worked on.

The MMC64 cannot handle multi-file programs, since it does not emulate a floppy disk drive. Any program that must load further data from the flash will not work, although a .d64/.d71-mount plugin seeks to alleviate this issue by replacing the DOS file access routines to simulate the presence of a floppy drive.

User-configured features

MMC64 allows user-supplied plugins for additional functionality not available in the BIOS. Many plugins already exist, including picture displayers for various Commodore pictureformats, .wav/.raw audio players, a fast .d64 writer plugin, an alternative SID player, an animation player for .ani files, viewers for ASCII text files, a reader for .t64 tape images, a launcher for .crt cartridge files (only those that do not employ custom ASICs, e.g. games), a .d81 writer for 1581 disk images, and others.

Also available is a .d64/.d71 mount plugin that allows reading from a disk image residing on the flash media. The drawback is that a Retro Replay is needed as well. As with the IDE64 IDE interface for the Commodore 64, the program or game has to use standard KERNAL file access routines and not rely on a software fast loader, which unfortunately is the norm, since the Kernel-routines are very slow when accessing a normal 1541. Games and programs fixed for the IDE64 usually work on MMC64 as well using this plugin.

Reading and writing from/to a disk image on the flash media so far is only possible with the .dfi plugin. However, the .d64 needs to be converted to .dfi and the program itself has to be modified with Dreamload as its fast loader (version 2.7 or above). Dreamload is compatible with all kinds of regular disk drives as well as Commodore compatible harddrives, IDE64, and now also MMC64. A .dfi file is a more convenient and compatible container for a .d64 file.

The MMC64 has an autostart feature that allows to run an arbitrary program named 'BOOT.BIN' in the /SYSTEM64 directory (e.g. a selection menu or an alternative file browser) instead of the built-in file browser. Many users prefer the TNT file browser for its ability to display long filenames and browse/launch from within .d64 files. The autostart can be temporarily disabled by holding the Commodore key upon starting the computer or permanently from within the BIOS's configuration menu.

2.37.3 See also

- Commodore 64 peripherals

- 1541 Ultimate

2.37.4 External links

- Wiki

- Technical Information on the MMC64 with a list of plugins and compatible flash media brands (German)

2.38   MOS Technology 6510

Image of the internals of a Commodore 64 showing the 6510 CPU (40-pin DIP, lower left). The chip on the right is the 6581 SID. The production week/year (WWYY) of each chip is given below its name.

The **MOS Technology 6510** is an 8-bit microprocessor designed by MOS Technology, Inc., and is a modified form of the very successful 6502.

The primary change from the 6502 was the addition of an 8-bit general purpose I/O port (only six I/O pins were available in the most common version of the 6510). In addition, the address bus could be made tristate.

The 6510 was only widely used in the Commodore 64 home computer and its variants. In the C64 the extra I/O pins of the processor were used to control the computer's memory

map by bank switching, and in the C64 also for controlling three of the four signal lines of the Datassette tape recorder (the electric motor control, key-press sensing and write data lines; the read data line went to another I/O chip). It was possible, by writing the correct bit pattern to the processor at address $01, to completely expose almost the full 64 KB of RAM in the C64, leaving no ROM or I/O hardware exposed except for the processor I/O port itself and its Data Directional Register.[1]

2.38.1 Variants

- **VCC:** Power (Usually +5v)   - **GND:** Ground
- **D7...D0:** Data Bus   - **A15...A0:** Address Bus
- **P5...P0:** I/O Port   - **/HALT:** Suspend Processing
- **CLKIN:** Master Clock Input - **CLKOUT:** Master Clock Output
- **/NMI:** Nonmaskable Interrupt - **/IRQ:** Maskable Interrupt
- **R/W:** Read (High)/Write (Low) - **/RESET:** Reset
- **CE:** Chip Select

Pin configuration of the most common variation of the 6510 CPU (a mistake labels the /RDY pin as /HALT in this image)

MOS 8500

In 1985, MOS produced the **8500**, an HMOS version of the 6510. Other than the process change, it is virtually identical to the NMOS version of the 6510. The 8500 was originally designed for use in the modernised C64, the C64C. However in 1985, limited quantities of 8500s were found on older NMOS-based C64s. It finally made its official debut in 1987, appearing in a motherboard using the new 85xx HMOS chipset.

MOS 7501/8501

The **7501/8501** variant of the 6510 was introduced in 1984.[2] It was used in Commodore's C16, C116 and Plus/4 home computers, where its I/O port controlled not only the Datasette but also the CBM Bus interface. The main difference between 7501 and 8501 CPUs is that they were manufactured with different technologies: 7501 was manufactured with HMOS-1 and 8501 with HMOS-2.[3] The NMI signal are not available for MOS 7501 and MOS 8501.[4]

MOS 8502

The 2 MHz-capable 8502 variant was used in the Commodore 128. All these CPUs are opcode compatible (including undocumented opcodes).[5]

MOS 6510T

The Commodore 1551 disk drive used the **6510T**, a version of the 6510 with eight I/O lines. The NMI and RDY signals are not available.

2.38.2   See also

- Interrupts in 65xx processors

2.38.3   References

[1] http://www.atarimagazines.com/compute/issue32/112_1_COMMODORE_64_ARCHITECTURE.php

[2] http://plus4world.powweb.com/hardware/MOS_75018501 Hardware - MOS 7501/8501

[3] http://plus4world.powweb.com/hardware/MOS_75018501 Hardware - MOS 7501/8501

[4] The NMI and RDY signals are not available.

[5] http://www.oxyron.de/html/opcodes02.html

2.38.4   External links

- MOS 6510 datasheet (GIF format, zipped)

- MOS 6510 datasheet (PDF format)

- MOS 6510 datasheet (Nov. 1982, PDF format)

- Marat Fayzullin's emulator page (includes downloadable source code for 6502)

- A web server using a MOS 6510 computer aka C64 unmodified

2.39   MOS Technology 8502

MOS 8502 Microprocessor

The **MOS Technology 8502** was an 8-bit microprocessor designed by MOS Technology and used in the Commodore 128. Based on the MOS 6510 that was used in the Commodore 64, the 8502 added the ability to optionally run at double clock rate of Commodore 64 with some limitations.

Since the 40-column VIC-II display chip could not "steal" sufficient cycles when the CPU ran at double speed, video display in fast mode was available only with the 80-column VDC (unlike the VIC, which shares memory with the CPU, the VDC has its own dedicated video RAM in the C128). Some 40-column applications selectively disabled the screen when performing CPU-intensive calculations so that the additional speed could be utilized when the loss of video output was unimportant. A smaller speed gain, about 35%, was also possible while keeping the 40-column display active, by switching to 2 MHz only while the VIC-II is drawing the vertical screen border, since no RAM access by the VIC is needed during that time.

The pinout is a little bit different from the 6510. The 8502 has an extra I/O-pin (the built-in I/O port mapped to addresses 0 and 1 is extended from 6 to 7 bits) and lacks the φ2-pin that the 6510 had.

2.39.1   Variants

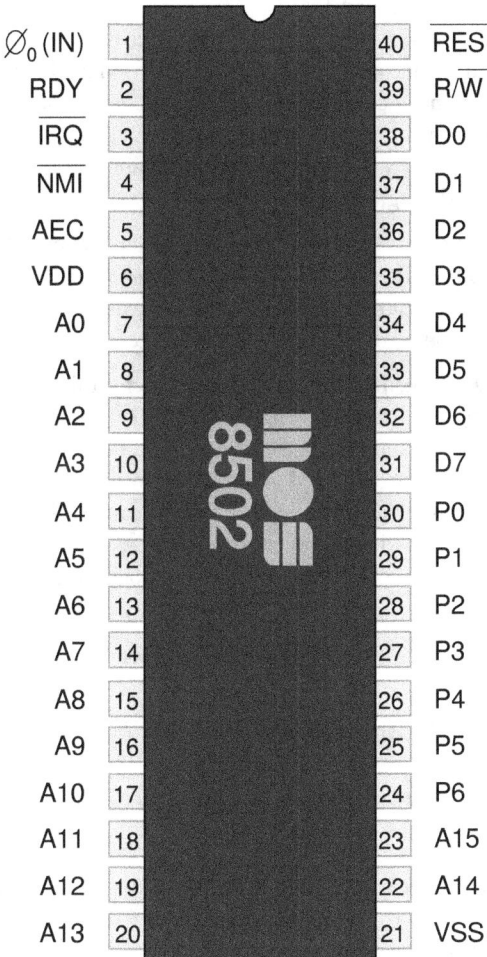

8502 pin configuration[1] (40-Pin DIP)

Ricoh 2A03

Ricoh 2A03 is used in Nintendo Entertainment System[2]

WDC 65SC02

WDC 65SC02 is used in Atari Lynx[3]

Ricoh 5A22

The Ricoh 5A22 is a microprocessor produced by Ricoh for the Super Nintendo Entertainment System (SNES) video game console. The 5A22 is based on the 16-bit CMD/GTE 65c816, itself a version of the WDC 65C816 (used in the Apple IIGS personal computer)..[4]

Sunplus SPLB31A

In 2007, HP released the HP 35s, a calculator that uses the Sunplus Technology / Generalplus SPLB31A/GPLB31A, an embedded chip integrating a 8502 microprocessor core alongside LC display and I/O controllers.[5] The HP 17bII+, the HP 12c Prestige, as well as a revised version of the HP 12c Platinum and the HP 12c Platinum 25th Anniversary Edition, all manufactured by Kinpo Electronics, are also based on this chip.

2.39.2   References

[1] *Service Manual C-128/C128D Computer*, Commodore Business Machines, PN-314001-08, November 1987

[2] http://docs2.minhateca.com.br/767372,BR,0,0,m6502.txt

[3] http://docs2.minhateca.com.br/767372,BR,0,0,m6502.txt

[4] http://docs2.minhateca.com.br/767372,BR,0,0,m6502.txt

[5] Richard Nass: *Tear Down: Scientific calculator boils design down to two ICs*, Embedded.com, January 2008

2.40   MOS Technology 8563

The **8563 Video Display Controller** (**VDC**) was an integrated circuit produced by MOS Technology. It was used in the Commodore 128 computer to generate an 80-column (640×200 pixel) RGB video display, running alongside a VIC-II which supported Commodore 64-compatible graphics. The DCR models (as well as a few D-models) of the C128 used the later and more technically advanced 8568 [D]VDC controller.

2.40.1   History and characteristics

Originally intended for a planned (but unreleased) UNIX-based business computer, Commodore designed the VDC into several prototype machines. Of these, only the Commodore 128 ever saw production. Unlike earlier MOS video chips such as the popular VIC-II, the VDC had dedicated video memory, 16 kilobytes (upgradable to 64 kilobytes) in the original or "flat" C128 and 64 kilobytes in the C128DCR. This RAM was not directly accessible by the microprocessor.

The 8563 was more difficult to produce than most of the rest of the MOS Technology line, and initial yields were very low. Also, there were timing issues with the VDC that would cause indirect load and store operations on its registers to malfunction.

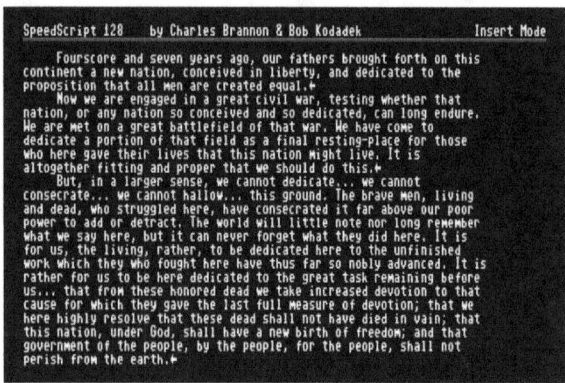

The VDC was designed with office suite applications in mind. Shown here is SpeedScript 128, *a word processor.*

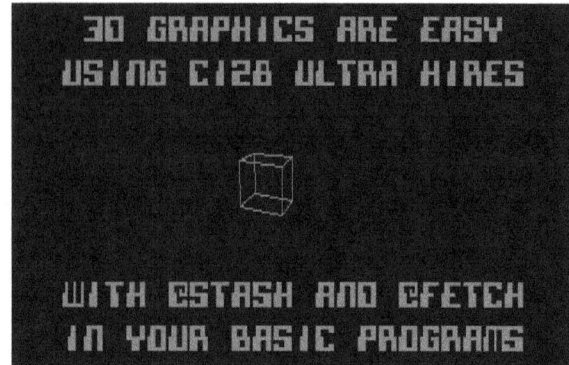

This Ultra Hi-Res *demo showcases the VDC's blitter capabilities with a simple 3D animation of a wire frame model of a cube.*

Officially, the VDC was a text-only chip, although a careful reading of the technical literature by MOS Technology that was given to the early C128 developers did indicate that a high-resolution bitmap mode was possible—it simply wasn't described in any detail. BASIC 7.0, the Commodore 128's built-in programming language, only supported high-resolution graphics in 40-column mode via the legacy VIC-II chip.

Shortly after the release of the C128 the VDC's bitmap mode was described in considerable detail in the Data Becker book "Commodore 128 - Das große GRAFIK-Buch" (published in late 1985 in the USA by Abacus Software), and an assembly language program was provided by the German authors Klaus Löffelmann and Dieter Vüllers, in which it was possible to set or clear any pixel or, using BASIC to perform the necessary calculations, generate bitmapped geometric shapes on the 80 column screen (Chapter 3.9.1 "VDC HI-RES-Grafik" Page 213ff). In February 1986, less than a year after the Commodore 128's release, *RUN* magazine published "*Ultra Hi-Res Graphics*", an article describing the VDC's bitmapped mode and including a type-in program (written in 8502 assembly lan-

guage) that extended BASIC 7.0's capabilities to support 640×200 high-resolution graphics using the 8563. Authors Lou Wallace and David Darus later developed the Ultra Hi-Res utility into a commercial package, *BASIC 8*. One of the most popular third-party utilities for the C128, this offered more advanced VDC high-resolution capabilities to a wide audience of programmers.

Commodore finally offered complete official documentation on the VDC in the *Commodore 128 Programmer's Reference Guide*. VDC bitmap modes were used extensively in the C128 version of the GEOS operating system.

The VDC lacked sprite capabilities, which limited its use in gaming applications. However, it did contain blitting capabilities to autonomously perform small block memory copies within its dedicated video RAM. While the VDC is performing such a copy, the system CPU can continue running code, provided no other VDC accesses are attempted before the copy is finished. These functions were used by the C128's screen editor ROM to rapidly scroll or clear screen sections.

2.40.2 Technical specifications

- RGBI output (RGB plus Intensity) compatible with IBM's CGA video standard.

- 16 or 64 kilobyte address space for display, character shape and display attribute memory (dedicated, separate from system memory).

- Up to 720×700 pixel video resolution in interlaced mode (maximum with 64 kilobyte video ram) . Other image sizes are possible, depending on programmer's needs, such as 640×200 non-interlaced, 640×400 interlaced, etc.

- 80×25 characters text resolution (C128 kernel default); other sizes such as 80×50 or 40×25 are possible.

- 8 colors at 2 intensities.

2.40.3 Programming

Addressing the VDC's internal registers and dedicated video memory must be accomplished by indirect means. First the program must tell the VDC which of its 37 internal registers is to be accessed. Next the program must wait until the VDC is ready for the access, after which a read or write on the selected internal register may be performed. The following code is typical of a register read:

ldx #regnum ;VDC register to access stx $d600 ;write to control register loop bit $d600 ;check bit 7 of status register

bpl loop ;VDC not ready lda $d601 ;read from VDC register
...

The following code is typical of a register write operation:

ldx #regnum ;VDC register to write to stx $d600 ;write to control register loop bit $d600 ;check bit 7 of status register bpl loop ;VDC not ready sta $d601 ;write to VDC register
...

Owing to this somewhat cumbersome method of controlling the VDC, the maximum possible frame rate in bitmapped mode is generally too slow for arcade-style action video games, in which bit-intensive manipulation of the display is required.

2.40.4   Register Listing

This information was adapted from the *Commodore 128 Programmer's Reference Guide*[1]

2.40.5   References

[1] Commodore Capital, Inc., (1986). *Commodore 128 programmer's reference guide.* New York, NY: Bantam Books, Inc.

2.40.6   External links

- *Ultra Hi-Res* self-extracting archive - Volume I

- *Ultra Hi-Res* self-extracting archive - Volume II

- *C= Hacking* volume 2 - Register listing and description

2.41   MOS Technology 8568

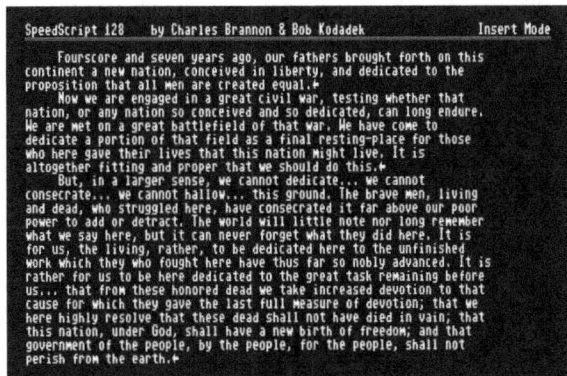

The VDC was designed with office suite applications in mind. Shown here is SpeedScript 128, *a word processor.*

This Ultra Hi-Res *demo showcases the VDC's blitter capabilities with a simple 3D animation of a wire frame model of a cube.*

The **8568 Video Display Controller (VDC)** was MOS Technology's graphics processor responsible for the 80 column or RGBI display on D[CR] models of the Commodore 128 personal computer. In the Commodore 128 service manual, this part was referred to as the "80 column CRT controller." The 8568 embodied many of the features of the older 6545E monochrome CRT controller plus RGBI color.

The original ("flat") C128 used the 8563 video controller to generate the 80 column display. The 8568 was essentially an updated version of the 8563, combining the latter's functionality with glue logic that previously was implemented by discrete components in physical proximity to the 8563. Unlike the 8563, the 8568 included an unused active low interrupt request line (/INTR), which was asserted when the "ready" bit in the 8568's status register changed from 0 to 1. Reading the control register would automatically de-assert /INTR. Owing to differences in pin assignments and circuit interfacing, the 8563 and 8568 are not electrically interchangeable.

The Commodore 128 had two video display modes, which were usually used singularly, but could be used simultaneously if the computer was connected to two compatible video monitors. The VIC-II chip, also found in the Commodore 64, was mapped directly into main memory—the video memory and CPUs (the 8502 and Z80A processors) shared a common 128 KB RAM, and the VIC-II control registers were accessed as memory locations (that is, they were memory mapped).

Unlike the VIC-II, the 8568 had its own local video RAM, 64K in the C-128DCR model (sold in North America) and, depending on the date of manufacture of the particular machine, either 16 or 64K in the C-128D model (marketed in Europe). Addressing the VDC's internal registers and dedicated video memory must be accomplished by indirect means. First the program must tell the VDC which of its 37 internal registers is to be accessed. Next the program must wait until the VDC is ready for the access, after which a read or write on the selected internal register may be performed. The following code is typical of a register read:

ldx #regnum ;VDC register to access stx $d600 ;write to control register loop bit $d600 ;check bit 7 of status register bpl loop ;VDC not ready lda $d601 ;read from VDC register ...

The following code is typical of a register write operation:

ldx #regnum ;VDC register to write to stx $d600 ;write to control register loop bit $d600 ;check bit 7 of status register bpl loop ;VDC not ready sta $d601 ;write to VDC register ...

Owing to this somewhat cumbersome method of controlling the 8568, the maximum possible frame rate in bit-mapped mode is generally too slow for arcade-style action video games, in which bit-intensive manipulation of the display is required.

The final versions of the 8568 had the revision codes R9a or R9b appended to the part number, apparently indicating undocumented improvements.

2.41.1 Features

- 80 × 25 characters text resolution

- 720 × 700 pixels maximum video resolution[1]

- Interlaced up to 80 × 50 text, 640H × 480V bitmap

- 3 character modes: standard, semigraphic and graphic, double width & HiRes bitmap.

- Output: digital RGBI with 16 colors or 16 gray shades, plus limited monochrome composite.

- Features: Interlace mode, horizontal & vertical scrolling, Light pen input, hardware cursor, underline, blink, reverse video, 2 character sets of 256 each, update ready interrupt

- Can access 64 KByte of memory, programmable to interface either 4164/4464 or 4416 DRAM

- 48 pins, +5 Volt DC supply.[2]

2.41.2 Register Listing

This information was adapted from the *Commodore 128 Programmer's Reference Guide*[3]

2.41.3 Notes

1. ^ In Commodore 128 terminology, the VIC-II display was called the 40 column display, and the VDC, 80 column, due to the number of columns of fixed-pitch text that could be natively displayed.

2. ^ Commodore service manual 314001-08 (1987).

3. ^ The 8563/8568 hardware is always visible on the address and data buses regardless of which mode in which the C-128 is operating. Hence it is possible to generate an 80 column display while running in C-64 compatibility mode. There is, of course, no C-64 operating system support.

2.41.4 References

[1] "empty". Graphic Booster 128

[2] "empty". Archived from the original on 2009-10-28. 090425 geocities.com

[3] Commodore Capital, Inc., (1986). *Commodore 128 programmer's reference guide.* p.294, New York, NY: Bantam Books, Inc.

[4] "Uncovered: The VDC 8568's 38th register". Archived from the original on 2011-08-15. Retrieved 2008-12-30.

2.42 MOS Technology CIA

The **6526/8521 Complex Interface Adapter** (**CIA**) was an integrated circuit made by MOS Technology. It served as an I/O port controller for the 6502 family of microprocessors, providing for parallel and serial I/O capabilities as well as timers and a Time-of-Day (TOD) clock. The device's most prominent use was in the Commodore 64 and Commodore 128(D), each of which included two CIA chips. The Commodore 1570 and Commodore 1571 floppy disk drives contained one CIA each. The Amiga home computers employed two and Commodore 1581 floppy disk drive one 8520 chip, which is functionally equivalent to 6526/8521 except the simplified TOD circuitry.

2.42.1 Parallel I/O

The CIA had two 8-bit bidirectional parallel I/O ports. Each port had a corresponding Data Direction Register, which allowed each data line to be individually set to input or output mode. A read of these ports always returned the status of the individual lines, regardless of the data direction that had been set.

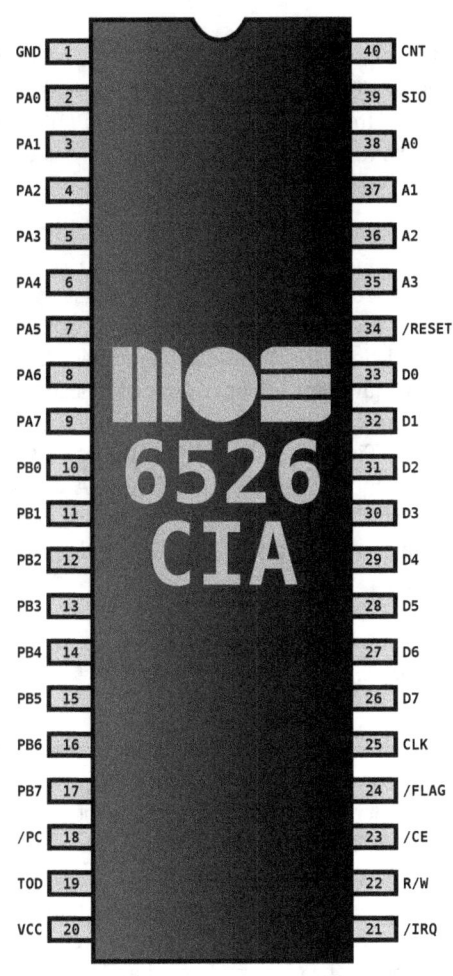

Pin configuration of the 6526 CIA

2.42.2   Serial I/O

An internal bidirectional 8-bit shift register enabled the CIA to handle serial I/O. The chip could accept serial input clocked from an external source, and could send serial output clocked with one of the built-in programmable timers. An interrupt was generated whenever an 8-bit serial transfer had completed. It was possible to implement a simple "network" by connecting the shift register and clock outputs of several computers together. The maximum bitrate is 500 kbit/s for the 2 MHz version.

2.42.3   Handshaking

Two dedicated control lines (/FLAG and /PC) were implemented to allow coordination between multiple CIA chips. These lines, along with 8 of the 16 available parallel port data lines, made it possible to use the CIA as a simple, Centronics-compatible line driver.

2.42.4   Interval timers

Two programmable interval timers were available, each with sub-microsecond precision. Each timer consisted of a 16-bit read-only presettable down counter and a corresponding 16-bit write-only latch. Whenever a timer was started, the timer's latch was automatically copied into its counter, and the counter would then decrement with each clock cycle until underflow, at which an interrupt would be generated.

The timer could run in either "one-shot" mode, halting after the first interrupt, or "continuous" mode, reloading the latch value again and starting the timer cycle anew. In addition to generating interrupts, the timer output could also be gated to the second I/O port.

As configured in the Commodore 64 and Commodore 128, the CIA's timing was controlled by the phase two system clock, nominally one MHz. This meant that the timers decremented at approximately one microsecond intervals, the exact time period being determined by whether the system used the NTSC or PAL video standard. In the C-128, clock stretching was employed so the CIA's timing was unaffected by whether the system was running in SLOW or FAST mode.

It was possible to generate relatively long timing intervals by programming timer B to count timer A underflows. If both timers were loaded with the maximum interval value of 65,535, a timing interval of one hour, 11 minutes, 34 seconds would result.

2.42.5   Time-of-Day (TOD) Clock

A real-time clock is incorporated in the CIA, providing a timekeeping device more conducive to human needs than the microsecond precision of the interval timers. Time is kept in the American 12 hour AM/PM format. The TOD clock consists of four read/write registers: hours (with bit 7 acting as the AM/PM flag), minutes, seconds and tenths of a second. All registers read out in BCD format, thus simplifying the encoding/decoding process.

Reading from the registers will always return the time of day. In order to avoid a carry error while fetching the time, reading the hours register will immediately halt register updating, with no effect on internal timekeeping accuracy. Once the tenths register has been read, updating will resume. It is possible to read any register other than the hours register "on the fly," making the use of a running TOD clock as a timer a practical application. If the hours register is read, however, it is essential to subsequently read the tenths register. Otherwise, all TOD registers will remain "frozen."

Setting the time involves writing the appropriate BCD values into the registers. A write access to the hours register will completely halt the clock. The clock will not start again until a value has been written into the tenths register. Owing to the order in which the registers appear in the system's memory map, a simple loop is all that is required to write the registers in the correct order. It is permissible to write to only the tenths register to "nudge" the clock into action, in which following a hardware reset, the clock will start at 1:00:00.0.

In addition to its timekeeping features, the TOD can be configured to act as an alarm clock, by arranging for it to generate an interrupt request at any desired time. Due to a bug in many 6526s (see also errata below), the alarm IRQ would not always occur when the seconds component of the alarm time is exactly zero. The workaround is to set the alarm's tenths value to 0.1 seconds.

The TOD clock's internal circuitry is designed to be driven by either 50 or 60 Hz clock signal, which can be inexpensively derived from the mains power source AC, resulting in a stable timekeeper with little long-term drift. The ability to work with both power line frequencies allowed a single version of the 6526 to be used in computers operated in countries with either 50 or 60 Hz mains power lines. It is important to note that contrary to the popular belief, NTSC or PAL video standards are not directly linked to mains power frequency. Additionally, some computers did not derive their TOD clock frequency from the mains power source. For example both NTSC and PAL variants of Commodore_SX-64 use 60Hz TOD clock supplied by a dedicated crystal. KERNAL operating system in Commodore_64 for example will determine the video standard during system startup, but tries neither to identify the supplied TOD clock frequency nor to initialise the CIAs correctly on 50Hz driven machines. Thus, it is the responsibility of any application software that wants to use either CIA's TOD function to determine the supplied frequency and set the CIA(s) flag accordingly itself. Failure to do so may cause the clock to deviate quickly from the correct time.

The 8520 revision of the CIA, as used in the Amiga and the Commodore 1581 disk drive, modified the time-of-day clock to be a 24-bit binary counter, replacing the BCD format of the 6526. Other behavior was similar, however.

2.42.6 Versions

The CIA was available in 1 MHz (6526), 2 MHz (6526A) and 3 MHz (6526B) versions. The form factor was a JEDEC-standard 40-pin ceramic or plastic DIP. The 8520 CIA, with its modified time-of-day clock, was used in the Amiga computers.

Commodore embedded reduced (just 4 registers) CIA-like logic for the cost reduced Commodore 1571 inside the C128DCR (See Commodore 128) in a gate array called 5710 which also contains other functions. The 5710 CIA has the serial clock for the fast serial interface hardwired to a CIA6526 equivalent Timer A value of 5, leading to a per-bit time of 5μs on transmission. This is different from what used to be a Timer A value of 6 in the 6526 CIA in the original Commodore 1571. The 5710 CIA does not contain timer or timer control registers. It only contains two port registers and the register to control the serial shifter and its event.

2.42.7 Errata

In addition to the aforementioned alarm clock interrupt bug, many CIAs exhibited a defect in which the part would fail to generate a timer B hardware interrupt if the interrupt control register (ICR) was read one or two clock cycles before the time when the interrupt should have actually occurred. This defect, as well as logic errors in the Commodore provided (8 bit) operating system, caused frequent pseudo-RS-232 errors in the Commodore 64 and Commodore 128 computers when running at higher baud rates.

2.42.8 External links

- MOS 6526 CIA datasheet (GIF format, zipped)

- MOS 6526 CIA datasheet (PDF format)

2.43 MOS Technology SID

The MOS Technology 6581/8580 **SID (Sound Interface Device)** is the built-in Programmable Sound Generator chip of Commodore's CBM-II, Commodore 64, Commodore 128 and Commodore MAX Machine home computers. It was one of the first sound chips of its kind to be included in a home computer prior to the digital sound revolution.

Together with the VIC-II graphics chip, the SID was instrumental in making the C64 the best-selling computer in history, and is partly credited for initiating the demoscene.

MOS Technology SIDs. The right chip is a 6581 from MOS Technology, known at the time as the Commodore Semiconductor Group (CSG.) The left chip is an 8580, also from MOS Technology. The numbers 0488 and 3290 are in WWYY form, i.e. the chips were produced week 4 1988 and week 32 1990. The last number is assumed to be a batch number.

The SID has U.S. Patent 4,677,890, which was filed on February 27, 1983, and issued on July 7, 1987. The patent expired on July 7, 2004.

2.43.1   Design process

The SID was devised by engineer Robert "Bob" Yannes, who later co-founded the Ensoniq digital synthesizer company. Yannes headed a team that included himself, two technicians and a CAD operator, who designed and completed the chip in five months, in the latter half of 1981. Yannes was inspired by previous work in the synthesizer industry and was not impressed by the current state of computer sound chips. Instead, he wanted a high-quality instrument chip, which is the reason why the SID has features like the envelope generator, previously not found in home computer sound chips.[1][2]

> I thought the sound chips on the market, including those in the Atari computers, were primitive and obviously had been designed by people who knew nothing about music.[2]
> — Robert Yannes, *On the Edge: The Spectacular Rise and Fall of Commodore*

Emphasis during chip design was on high-precision frequency control, and the SID was originally designed to have 32 independent voices, sharing a common oscillator.[2] However these features could not be finished in time, so instead the mask work for a certain working oscillator was simply replicated three times across the chip's surface, creating three voices each with its own oscillator. Another feature that was not incorporated in the final design was a frequency look-up table for the most common musical notes, a feature that was dropped because of space limitations.[3]

The support for an audio input pin was a feature Yannes added without asking, even though this had no practical use in a computer, although it enabled the chip to be used as a simple effect processor. The masks were produced in 7-micrometer technology to gain a high yield; the state of the art at the time was 6-micrometer technologies.[3]

The chip, like the first product using it (the Commodore 64), was finished in time for the Consumer Electronics Show in the first weekend of January 1982. Even though Yannes was partly displeased with the result, his colleague Charles Winterble said: "This thing is already 10 times better than anything out there and 20 times better than it needs to be."[4]

The specifications for the chip were not used as a blueprint. Rather, they were written as the development work progressed, and not all planned features made it into the final product. Yannes claims he had a feature-list of which three quarters made it into the final design. This is the reason why some of the specifications for the first version (6581) were accidentally incorrect. The later revision (8580) was revised to match the specification. For example, the 8580 expanded on the ability to perform a logical AND between two waveforms, something that the 6581 could only do in a somewhat limited and unintuitive manner. Another feature that differs between the two revisions is the filter: the 6581 version is far away from the specification.

2.43.2   Manufacturing, remarking, and forgery

Since 6581 and 8580 SID ICs are no longer produced, they have become highly sought after. In late 2007, various defective chips started appearing on eBay as supposedly "new".[5] All of these remarked SIDs have a defective filter, but some also have defective channels/noise generators, and some are completely dead. The remarked chips are assumed to either be factory rejects from back when the chip was still produced, or possibly 'reject pulls' from one of the chip pulling operations which were used to supply the chips used in the Elektron SIDStation and the HardSID cards. Fake SID chips have also been supplied to unwitting buyers from unscrupulous manufacturers in China; the supplied chips are laser-etched with completely bogus markings, and the chip inside the package is not a SID at all.[6]

2.43.3   Features

- three separately programmable independent audio oscillators (8 octave range, approximately 16 - 4000 Hz)

- four different waveforms per audio oscillator

(sawtooth, triangle, pulse, noise)

- one multi mode filter featuring low-pass, high-pass and band-pass outputs with 6 dB/oct (bandpass) or 12 dB/octave (lowpass/highpass) rolloff. The different filter modes are sometimes combined to produce additional timbres, for instance a notch-reject filter.

- three attack/decay/sustain/release (ADSR) volume controls, one for each audio oscillator.

- three ring modulators.

- oscillator sync for each audio oscillator.

- two 8-bit A/D converters (typically used for game control paddles, but later also used for a mouse)

- external audio input (for sound mixing with external signal sources)

- random number/modulation generator

2.43.4 Technical details

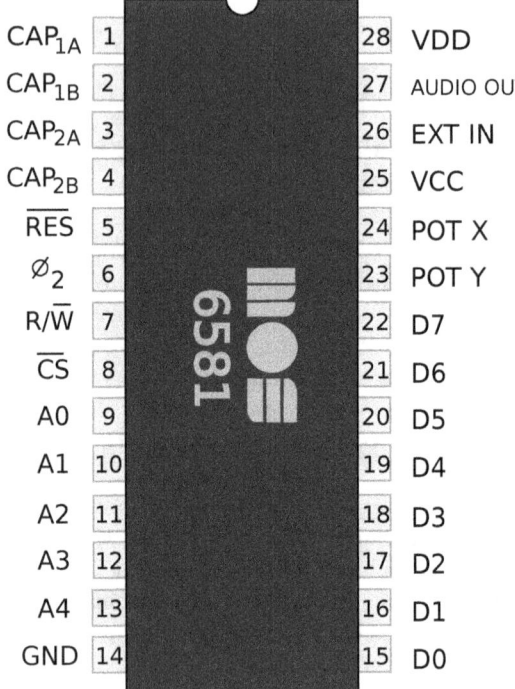

6581/6582/8580R5 Pin configuration

The SID is a mixed-signal integrated circuit, featuring both digital and analog circuitry. All control ports are digital, while the output ports are analog. The SID features three-voice synthesis, where each voice may use one of at least

five different waveforms: pulse wave (with variable duty cycle), triangle wave, sawtooth wave, pseudorandom noise (called white noise in documentation), and certain complex/combined waveforms when multiple waveforms are selected simultaneously. A voice playing Triangle waveform may be ring-modulated with one of the other voices, where the triangle waveform's bits are inverted when the modulating voice's msb is set, producing a discontinuity and change of direction with the Triangle's ramp. Oscillators may also be hard-synced to each other, where the synced oscillator is reset whenever the syncing oscillator's msb raises.

Each voice may be routed into a common, digitally controlled analog 12 dB/octave multimode filter, which is constructed with aid of external capacitors to the chip. The filter has lowpass, bandpass and highpass outputs, which can be individually selected for final output amplification via master volume register. Using a combined state of lowpass and highpass results in a notch (or inverted bandpass) output.[7] The programmer may vary the filter's cut-off frequency and resonance. An external audio-in port enables external audio to be passed through the filter.

The ring modulation, filter, and programming techniques such as arpeggio (rapid cycling between 2 or more frequencies to make chord-like sounds) together produce the characteristic feel of SID music.

Due to imperfect manufacturing technologies of the time and poor separation between the analog and digital parts of the chip, the 6581's output (before the amplifier stage) was always slightly biased from the zero level. By adjusting the amplifier's gain through the main 4-bit volume register, this bias could be modulated as PCM, resulting in a "virtual" fourth channel allowing 4-bit digital sample playback. The glitch was known and used from an early point on, first by Electronic Speech Systems to produce sampled speech in games such as Impossible Mission (1983, Epyx) and Ghostbusters (1984, Activision). The first instance of samples being used in actual musical compositions was by Martin Galway in Arkanoid (1987, Imagine), although he had copied the idea from an earlier drum synthesizer package called Digidrums. The length of sampled sound playback was limited first by memory and later technique. Kung Fu Fighting (1986), a popular early sample, has a playback length measured in seconds. c64mp3 (2010) and Cubase64 (2010) demonstrate playback lengths measured in minutes. Also, it was hugely CPU intensive - one had to output the samples very fast (in comparison to the speed of the 6510 CPU).

The better manufacturing technology in the 8580 used in the later revisions of Commodore 64C and the Commodore 128DCR caused the bias to almost entirely disappear, causing the digitized sound samples to become very quiet. Fortunately, the volume level could be mostly restored with either a hardware modification (biasing the audio-in pin), or more commonly a software trick involving using the Pulse waveform to intentionally recreate the required bias. The software trick generally renders one voice temporarily unusable, although clever musical compositions can make this problem less noticeable. An excellent example of this quality improvement noticeably reducing a sampled channel can be found in the introduction to Electronic Arts' game Skate or Die (1987). The guitar riff played is all but missing when played on the Commodore 64c or the Commodore 128.

At the X'2008 demo party, a completely new method of playing digitized samples was unveiled. The method allows for an unprecedented four (software-mixed) channels of 8-bit samples with optional filtering on top of all samples, as well as two ordinary SID sound channels.[8][9] The method works by resetting the oscillator using the waveform generator test bit, quickly ramping up the new waveform with the Triangle waveform selected, and then disabling all waveforms, resulting in the DAC continuing to output the last value---which is the desired sample. This continues for as long as two scanlines, which is ample time for glitch-free, arbitrary sample output. It is however more CPU-intensive than the 4-bit volume register DAC trick described above. Because the filtering in a SID chip is applied after the waveform generators, samples produced this way can be filtered normally.

The original manual for the SID mentions that if several waveforms are enabled at the same time, the result will be a binary AND between them. What happens in reality is that the input to the waveform DAC pins receive several waveforms at once. For instance, the Triangle waveform is made with a separate XOR circuit and a shift-to-left circuit. The top bit drives whether the XOR circuit inverts the accumulator value seen by the DAC. Thus, enabling triangle and sawtooth simultaneously causes adjacent accumulator bits in the DAC input to mix. (The XOR circuit does not come to play because it is always disabled whenever the sawtooth waveform is selected.) The pulse waveform is built by joining all the DAC bits together via a long strip of polysilicon, connected to the pulse control logic that digitally compares current accumulator value to the pulse width value. Thus, selecting the pulse waveform together with any other waveform causes every bit on the DAC to partially mix, and the loudness of the waveform is affected by the state of the pulse.

The noise generator is implemented as a 23-bit-length linear feedback shift register (Feedback polynomial: $x^{22}+x^{17}+1$).[10][11] When using noise waveform simultaneously with any other waveform, the pull-down via waveform selector tends to quickly reduce the XOR shift register to 0 for all bits that are connected to the output DAC. As the zeroes shift in the register when the noise is clocked, and no 1-bits are produced to replace them, a situation can arise where the XOR shift register becomes fully zeroed. Luckily, the situation can be remedied by using the waveform control test bit, which in that condition injects one 1-bit into the XOR shift register. Some musicians are also known to use noise's combined waveforms and test bit to construct unusual sounds.

The 6581 and 8580 differ from each other in several ways. The original 6581 was manufactured using the older NMOS process, which used 12V DC to operate. The 6581 is very sensitive to static discharge and if they weren't handled properly the filters would stop working, explaining the reason of the great quantity of dead 6581s in the market. The 8580 was made using the HMOS-II process, which requires less power (9V DC), and therefore makes the IC run cooler. The 8580 is thus far more durable than the 6581. Also, due to stabler waveform generators, the bit-mixing effects are less noticeable and thus the combined waveforms come close to matching the original SID specification (which stated that they will be combined as a binary AND). The filter is also very different between the two models, with the 6581 cutoff range being a relatively straight line on a log scale, while the cutoff range on the 8580 is a straight line on a linear scale, and is close to the designers' actual specifications. Additionally, a better separation between the analog and the digital circuits made the 8580's output less noisy and distorted. The noise in 6xxx-series systems can be reduced by disconnecting the audio-in pin.

The consumer version of the 8580 was rebadged the 6582, even though the die on the chip is identical to a stock 8580 chip, including the '8580R5' mark. Dr. Evil Laboratories used it in their SID Symphony expansion cartridge (sold to Creative Micro Designs in 1991), and it was used in a few other places as well, including one PC sound-card.

Despite its documented shortcomings, many SID musicians prefer the flawed 6581 chip over the corrected 8580 chip. The main reason for this is that the filter produces strong distortion that is sometimes used to produce simulation of instruments such as a distorted electric guitar. Also, the highpass component of the filter was mixed in 3 dB attenuated compared to the other outputs, making the sound more bassy. In addition to nonlinearities in filter, the D/A circuitry used in the waveform generators produces yet more additional distortion that made its sound richer in character.

Revisions

No instances reading "6581 R1" ever reached the market. In fact, Yannes has stated that "[the] SID chip came out pretty well the first time, it made sound. Everything we needed for the show was working after the second pass." High-resolution photos of Charles Winterble's prototype

6581R1 produced in 1982

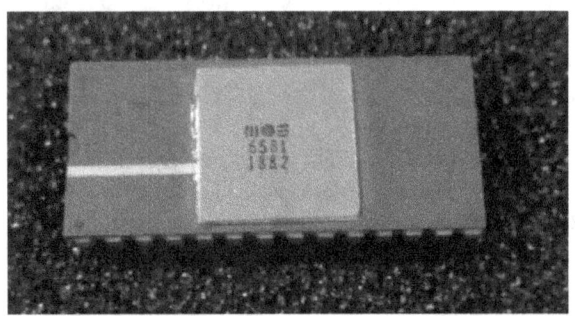

6581 produced in 1982

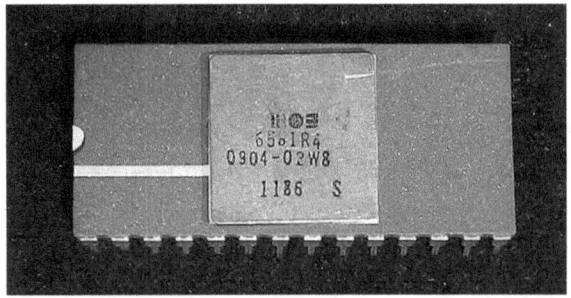

6581R4 CDIP produced in 1986

6582 produced in 1986

6582A produced in 1989

6582A produced in 1992

8580R5 produced 1986 in the U.S.

C64 show the markings "MOS 6581 2082", the last number being a date code indicating that his prototype SID chip was produced during the 20th week of 1982, which would be within 6 days of May 17, 1982.

These are the known revisions of the various SID chips:

(datecodes are in WWYY w=week y=year format)

- 6581 R1 - Prototype, only appeared on the CES machines and development prototypes, has a datecode of 4981 to 0882 or so. Has the full 12 bit filter cutoff range. An unknown number were produced, probably between 50 and 100 chips. All are ceramic packages.

- 6581 R2 - Will say "6581" only on the package. Filter

cutoff range was reduced to 11 bits and the MSB bit disconnected/forced permanently on, but is still on the die.The filter is leaky at some ranges and they tend to run hotter than other sid revisions. Made from 1182 until at least 1483. First 10 weeks or so of chips have ceramic packages (these usually appear on engineering prototypes but a few are on sold machines), the rest have plastic packages.

- 6581 R3 - Will say "6581" only, "6581 R3" or "6581 CBM" on the package. Had a minor change to the protection/buffering of the input pins. No changes were made to the filter section. Made from before 2083 until 1386 or so. The 6581R3 since around the week 47 of 1985 made in the Phillipines use the HMOS HC-30 degree silicon though the manufacturing process remained NMOS.

- 6581 R4 - Will say "6581 R4" on the package. Silicon grade changed to HMOS-II "HC-30" grade, though the manufacturing process for the chip remained NMOS. Produced from 4985 until at least 2590.

- 6581 R4 AR - Will say "6581 R4 AR" on the package. Minor adjustment to the silicon grade, no die change from R4. Produced from around 1986 (week 22) until at least the year 1992.

- 6582 - Will say "6582" on the package. Typically produced around the year 1986 in Hong Kong.

- 6582 A - Will say "6582A" (or "6582 A") on the package. Typically produced around the years 1989, 1990 and 1992 in the Philippines.

- 8580 R5 - Will say "8580R5" on the package. Produced from the years 1986 to 1993 in the Philippines, Hong Kong and in the US.

Some of these chips are marked "CSG" ("Commodore Semiconductor Group") and the Commodore Logo, while others are marked with "MOS". This includes chips produced during the same week (and thus, receiving the same date code), indicating that at least two different factory lines were in operation during that week. The markings of chips varied by factory and even by line within a factory throughout most of the manufacturing run of the chip.

2.43.5   Game audio

The majority of games produced for the Commodore 64 made use of the SID chip, with sounds ranging from simply clicks and beeps to complex musical extravaganzas or even entire digital audio tracks.

Well known composers of game music for this chip are Martin Galway, known for many titles, including *Wizball*, and Rob Hubbard, known for titles such as *ACE 2*, *Commando*, *Delta*, *International Karate*, *IK+*, and *Monty on the Run*. Other noteworthies include Jeroen Tel (*Cybernoid* and *Myth*), David Dunn (*Finders Keepers* and *Flight Path 737*) and Chris Hülsbeck, whose composition career started with the SID but has spanned nearly every kind of computer music and other synthesizers since.

2.43.6   Emulation

The fact that many enthusiasts prefer the real chip sound over software emulators has led to several recording projects aiming to preserve the authentic sound of the SID chip for modern hardware.

The sid.oth4 project[12] has over 380 songs of high quality MP3 available recorded on hardsid hardware and the SOASC= project[13] have the entire High Voltage SID Collection release 49 (over 35,000 songs) recorded in from real Commodore 64s in high quality mp3. Both projects emphasize the importance of preserving the authentic sound of the SID chip.

2.43.7   Software emulation

- In 1989 on the Amiga computer, the demo "The 100 Most Remembered C64 Tunes" and later the PlaySID application was released, developed by Per Håkan Sundell and Ron Birk. This was one of the first attempts to emulate the SID in software only, and also introduced the file format for representing songs made on the C64 using the SID chip. This later spawned the creation of similar applications for other platforms as well as the creation of a community of people fascinated by SID music, resulting in *The High Voltage SID Collection* which contains over 45,000 SID tunes.

A SID file contains the 6510 program code and associated data needed to replay the music on the SID. The SID files have the MIME media type audio/prs.sid.

The actual file format of a SID file has had several versions. The older standard is PSID (current version V3). The newer standard, RSID, is intended for music that requires a more complete emulation of the Commodore 64 hardware.

The SID file format is not a native format used on the Commodore 64 or 128, but a format specifically created for emulator-assisted music players such as *PlaySID* , *Sidplay* and JSidplay2. However, there are loaders like *RealSID-Play* and converters such as PSID64 that make it possible to play a substantial portion of SID files on original Commodore computers.

- SIDPlayer, developed by Christian Bauer and released in 1996 for the BeOS operating system, was the first SID emulator to replicate the filter section of the SID chip using a second-order Infinite impulse response filter as an approximation.

- In June 1998, a cycle-based SID emulator engine called reSID became available. The all-software emulator, available with C++ source code, is licensed under the GPL by the author, Dag Lem. In 2008, Antti Lankila significantly improved the filter and distortion simulation in reSID.[14] The improvements were included in VICE version 2.1 as well.

- In 2007 the JSidplay2 project was released, a pure Java based SID player developed by Ken Händel.

2.43.8   Hardware reimplementations

- In 2008 the HyperSID project is released. HyperSID is a VSTi which acts like a MIDI controller for HyperSID hardware unit (synthesizer based on SID chip) and developed by HyperSynth company.

Hardware Implementations using the SID chip

- In 1989 Innovation Computer developed the Innovation Sound Standard, an IBM PC compatible sound card with a SID chip and a game port. MicroProse promised software support for the card, and Commodore BASIC programs that used SID required little conversion to run on GW-BASIC.[15]

- In 1997, an electronic musical instrument utilizing the SID chip as its synthesis engine was released. It is called the SidStation, built around the 6581 model SID chip (as opposed to the newer 8580),[16] and it's produced by Swedish company Elektron. As the SID chip had been discontinued for years, Elektron allegedly bought up almost all of the remaining stock. In 2004, Elektron released the Monomachine pattern-based sequencer with optional keyboard. The Monomachine contains several synthesis engines, including an emulated 6581 oscillator using a DSP.

- In 1999 HardSID, another PC sound card, was released. The card uses from one to four SID chips and allows a PC to utilize the sound capabilities of the chip directly, instead of by emulation via generic sound cards (e.g. SoundBlaster).

- The Catweasel from German company Individual Computers, a PCI + Zorro multiformat floppy disk controller and digital joystick adapter for PCs, Macs, and Amigas, includes a hardware SID option, i.e. an option to insert one or two real SID chips in a socket for use when playing .MUS files.

- The MIDIbox SID is a MIDI-controlled synthesizer which can contain up to eight SID chips. It is a free open source project using a PIC microcontroller. Control of the synthesizer is realized with software or via a control panel with knobs, LEDs, LCD, etc., which may optionally be mounted on a keyboardless Commodore 64 body.

- The Prophet64 is a cartridge for the Commodore 64. It features four separate music applications, mimicking everything from modern sequencers to the Roland 303/909 series. With an optional User Port peripheral, the Prophet64 may synchronized to other equipment using DIN Sync standard (SYNC 24). The website now states "Prophet64 has been replaced with the MSSIAH."

- The MSSIAH is a cartridge for the Commodore 64 that replaces the Prophet64.

- Artist/Hacker Paul Slocum developed the Cynthcart cartridge that enables you to turn your C64 into an analogue synthesizer. The Cynthcart is available through atariage.com.

- The Parallel Port SID Interface allows those with very slim budgets to connect the SID chip to a PC.

- In May 2009 the SID chip was interfaced to the BBC Micro and BBC Master range of computers via the 1 MHz bus allowing music written for the SID chip on the Commodore 64 to be ported and played on the BBC Micro.

- In October 2009 thrashbarg's project interfaced an SID chip to an ATmega8 to play MIDI files on a MOS 6581 SID.

- In March 2010 STG published the SIDBlaster/USB - an open source, open hardware implementation of the SID that connects to (and is powered by) a USB port, using an FTDI chip for the USB interface and a PIC to interface the SID.

- In August 2010 SuperSoniqs published the Playsoniq, a cartridge for MSX computers, with (in addition to other features) a real SID on it, ready to use on any MSX machine.

SID hardware clones

- The SwinSID is hardware emulation of the SID using an Atmel AVR processor, also featuring a real SID player based on the Atmel AVR processor.

- The V-SID 1.0 project (code name SID 6581D, 'D' for digital) from David Amoros was born in 2005. This project is a hardware emulation of the SID chip from the Bob Yannes's interview, datasheets. The V-SID 1.0 engine had been implemented in a FPGA EP1C12 Cyclone from ALTERA, on an ALTIUM development board, and emulates all the characteristics of the original SID, except the filter which is a digital version (IIR filter controlled by a CPU).

- The PhoenixSID 65X81 project (2006) aimed to faithfully create the SID sound using modern hardware. The workings of a SID chip were recreated on an FPGA, based on interviews with the SID's creator, original datasheets, and comparisons with real SID chips. It was distinguished from similar attempts by its use of real analog circuitry instead of emulation for the legendary SID filter. However, the project was discontinued, because George Pantazopoulos, who was the head of this project, died on April 23, 2007, at the age of 29.

- The C64 Direct-to-TV emulates large portions the SID hardware, minus certain features such as (most notably) the filters. It reduces the entire C64 to a small circuit that fits into a joystick while sacrificing some compatibility.

2.43.9   Conventional music

SID sounds and snippets of SID music has been introduced into mainstream music at several occasions:

- In the spring of 1999 Zombie Nation released a remix of game musician David Whittaker's *Lazy Jones* (originally written for the SID in 1984) under the title *Kernkraft 400*. They used an Elektron SidStation for the sound.

- In 2000 8 Bit Weapon acquired a SidStation used on the track "Femmachine SID Mix" as well as for their work for Kraftwerk's "Space Lab" remix on astralwerks records. By 2009 they also added the MSSIAH c64 music workstation cart and began using the MOS 6581R4 & 8580 SID Chips in their music for remixes, TV scores, Video Game Scores, and their own album releases as well as the Sony released ACID loop library "8 Bit Weapon: A Chiptune Odyssey."

- In 2001 Bas Bron sampled the drums from Jeroen Tel's and Reyn Ouwehand's song made for the Rubicon game in the song *You've got my love*.

- In 2007 Timbaland's extensive use of the SidStation led to the 2007 Timbaland plagiarism controversy around his tracks *Block Party* and *Do It* (written for Nelly Furtado).

- SidStation is essential to the sound of Swedish band Machinae Supremacy. The band defines itself as SID metal.

- The Swedish acid ambient artist Carbon Based Lifeforms features a song called MOS 6581 on their 2003 album Hydroponic Garden.

2.43.10   See also

- Commodore 64 Games that Talk - A web page devoted to all the c64 games that contain Voice or Digitized speech, that demonstrates the SID chips ability to speak, as clearly as a real person.

- MOS Technology VIC - the combined graphics and sound chip of the VIC-20

- Atari POKEY

- MOS Technology 8364 "Paula"

- Chiptune

- Sound chip

- The High Voltage SID Collection

- Press Play on Tape, a C64 revival band

- Machinae Supremacy

- 8 Bit Weapon

2.43.11   References

[1] Perry, Tekla S.; Wallich, Paul (March 1985). "Design case history: the Commodore 64" (PDF). *IEEE Spectrum*: 48–58. ISSN 0018-9235. Retrieved 2011-11-12.

[2] Bagnall, Brian (2005). "The Secret Project 1981". *On the Edge - The Spectacular Rise and Fall of Commodore* (1 ed.). Winnipeg, Manitoba: Variant Press. p. 235. ISBN 0-9738649-0-7.

[3] Bagnall, Brian (2005). "The Secret Project 1981". *On the Edge - The Spectacular Rise and Fall of Commodore* (1 ed.). Winnipeg, Manitoba: Variant Press. p. 236. ISBN 0-9738649-0-7.

[4] Bagnall, Brian (2005). "The Secret Project 1981". *On the Edge - The Spectacular Rise and Fall of Commodore* (1 ed.). Winnipeg, Manitoba: Variant Press. p. 237. ISBN 0-9738649-0-7.

[5] http://kevtris.org/Projects/sid/remarked_sids.html re-marked SIDs

[6] http://www.lemon64.com/forum/viewtopic.php?t=27739. Retrieved 31 January 2014. Missing or empty |title= (help)

[7] "MIDIbox SID V2-User manual".

[8] "New revolutionary C64 music routine unveiled".

[9] "Mixer-SounDemon".

[10] "Sound Interface Device reference".

[11] "Examination of SID noise waveform".

[12] sid.oth4 project, sid.oth4 project

[13] SOASC= project, SOASC= project

[14] "JSIDPlay2: a cross-platform SID player and C64 emulator".

[15] Latimer, Joey (August 1989). "Innovation Sound Standard". *Compute!*. p. 68. Retrieved 11 November 2013.

[16] Revision as of 20:29, 19 May 2009 of "Commodore 64" wikipage

2.43.12 Notes

- Appendix O, "6581 Sound Interface Device (SID) Chip Specifications", of the *Commodore 64 Programmer's Reference Guide* (see the C64 article).

- Bagnall, Brian. *On The Edge: The Spectacular Rise and Fall of Commodore*, pp. 231–238,370–371. ISBN 0-9738649-0-7.

- Commodore 6581 Sound Interface Device (SID) datasheet. October, 1982.

2.43.13 Further reading

- Karen Collins, *"Loops and bloops". Music of the Commodore 64 games*, soundscapes, volume 8, Feb 2006

2.43.14 External links

SID information

- SID in-depth information page
- SID at DMOZ

- The 6581 SID Datasheet

- SID programming info

- MOS 8580 SID die shots

2.44 MOS Technology VIC-II

MOS 6569R3 (PAL version) on a C64 main board

The **VIC-II (Video Interface Chip II)**, specifically known as the MOS Technology 6567/8562/8564 (NTSC versions), 6569/8565/8566 (PAL), is the microchip tasked with generating Y/C video signals (combined to composite video in the RF modulator) and DRAM refresh signals in the Commodore 64 and C128 home computers.

Succeeding MOS's original VIC (used in the VIC-20), the VIC-II was one of the two chips mainly responsible for the C64's success (the other chip being the 6581 SID).

2.44.1 Development history

The VIC-II chip was designed primarily by Al Charpentier and Charles Winterble at MOS Technology, Inc. as a successor to the MOS Technology 6560 "VIC". The team at MOS Technology had previously failed to produce two graphics chips named *MOS Technology 6562* for the Commodore TOI computer, and *MOS Technology 6564* for the Color PET, due to memory speed constraints.[1]

In order to construct the VIC-II, Charpentier and Winterble made a market survey of current home computers and video games, listing up the current features, and what features they wanted to have in the VIC-II. The idea of adding sprites came from the Texas Instruments TI-99/4A computer and its TMS9918 graphics coprocessor. The idea to support collision detection came from the Mattel Intellivision. The Atari 800 was also mined for desired features.[2][3] About 3/4 of the chip surface is used for the sprite functionality.[4]

The chip was partly laid out using electronic design automation tools from *Applicon* (now a part of UGS Corp.), and partly laid out manually on vellum paper. The design was partly debugged by fabricating chips containing small subsets of the design, which could then be tested separately. This was easy since MOS Technology had both its research and development lab and semiconductor plant at the same location.[5] The chip was developed in 5 micrometer technology.[2]

The work on the VIC-II was completed in November 1981 while Robert Yannes was simultaneously working on the SID chip. Both chips, like the Commodore 64, were finished in time for the Consumer Electronics Show in the first weekend of January 1982.[6]

2.44.2   VIC-II features

- 16 kB address space for screen, character and sprite memory

- 320 × 200 pixels video resolution (160 × 200 in multicolor mode)

- 40 × 25 characters text resolution

- Three character display modes and two bitmap modes

- 16 colors

- Concurrent handling of 8 sprites per scanline, each of 24 × 21 pixels (12 × 21 multicolor)

- Raster interrupt (see details, below)

- Smooth scrolling

- Independent dynamic RAM refresh

- Bus mastering for a 6502-style system bus; CPU and VIC-II accessing the bus during alternating half-clock cycles (the VIC-II will halt the CPU when it needs extra cycles)

2.44.3   Technical details

Note that below register addresses are stated as seen by CPU in a C64. To yield the register numbers as usually given in data sheets (i. e. starting with 0), the leading "D0" should be omitted.

Programming

The VIC-II is programmed by manipulating its 47 control registers (up from 16 in the VIC), memory mapped to the range $D000–$D02E in the C64 address space. Of all these registers, 34 deal exclusively with sprite control (sprites being called MOBs, from "Movable Object Blocks", in the

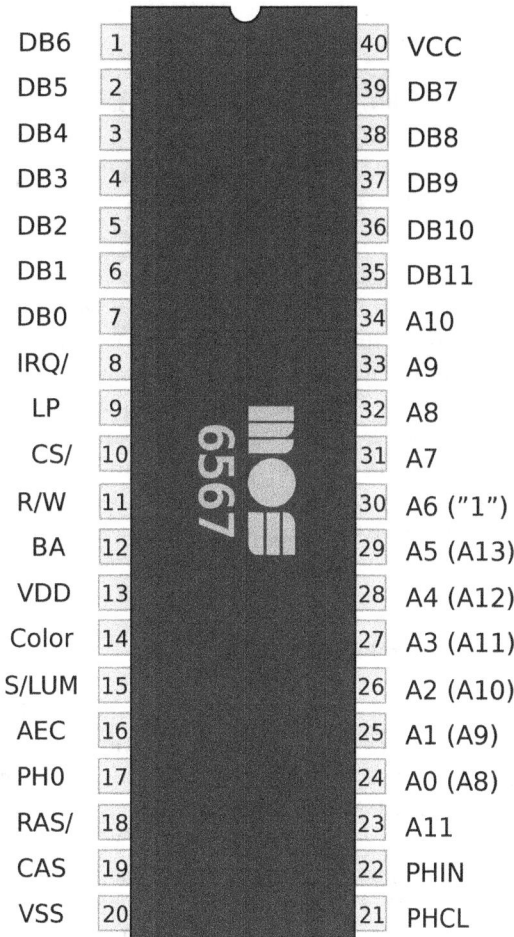

MOS 6567 VIC-II pinout.

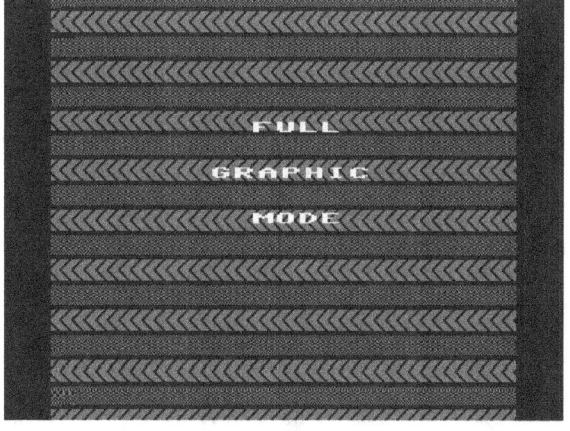

Supratechnic, *a type-in program published by* COMPUTE!'s Gazette *in November 1988, showcases the careful use of raster interrupts to display information outside of the standard screen borders (here: the upper and lower border).*

VIC-II documentation). Like its predecessor, the VIC-II handles light pen input, and with help from the C64's standard character ROM, provided the original PETSCII character set from 1977 on a similarly dimensioned display as the 40-column PET series.

By reloading the VIC-II's control registers via machine code hooked into the raster interrupt routine (the scanline interrupt), one can program the chip to generate significantly more than 8 concurrent sprites (a process known as sprite multiplexing), and generally give every program-defined slice of the screen different scrolling, resolution and color properties. The hardware limitation of 8 sprites per scanline could be increased further by letting the sprites flicker rapidly on and off. Mastery of the raster interrupt was essential in order to unleash the VIC-II's capabilities. Many demos and some later games would establish a fixed "lock-step" between the CPU and the VIC-II so that the VIC registers could be manipulated at exactly the right moment.

Character graphics

Most programming of the VIC-II is done with programmable character mode, and this is what the vast majority of C64/C128 games use. In power-on default mode, the character ROM is used which contains the PETSCII set. Normally, it can be seen only by the VIC-II and not the CPU. It is mapped into memory locations $3000–$3FFF and $B000-$BFFF and because of this, graphics data cannot be stored in those areas since the VIC-II will instead see the ROM there. By adjusting the bits in $01, the ROM can be mapped into $D000-$DFFF where it becomes visible to the CPU and programmers may copy characters from it to a different location as needed.

Up to 256 characters can be accessed by the VIC-II at once, although there is no limit to how many can be in memory provided they do not exceed the 16k video page. The default character set consists of two groups of 128 characters, the second group merely being an "inverse video" version of the first group.

Each character takes 8 bytes of memory to store. In addition to charsets, the VIC-II also uses 1k for its screen memory ($400–$7EF being the default). Color RAM is at $D800-$DBFF and cannot be moved from that location. It contains the values for Color 1 of each character.

In default hires character mode, the foreground of each character may be set individual per the color RAM. In multicolor character mode, Color 1 is limited to the first eight possible color values; the fourth bit is then used as a flag indicating if this character is to be displayed in hires or multicolor, thus making it possible to mix both types on one screen. Colors 2 and 3 are set by the registers at $D022 and $D023 and are global for all characters.

If Extended Background Color Mode is used, the upper two bits of the character code are used to select one of four background color registers. This allows four different background colors on the screen, but at the expense of only allowing 64 different characters instead of 256. Because this is fairly limiting, games seldom used it.

Bitmap mode

Adding an all-points-addressable bitmap mode was one of the Commodore design team's primary goals, as the VIC-I lacked such a feature. However, in order to use as little additional circuitry as possible, they organized it in the same manner as character mode, i.e. 8x8 and 4x8 tiles. Bitmap graphics require an 8k page for the pixel data and each byte corresponds to one row of eight or four pixels. The next byte is the row underneath it and after the 8th row, returning to the top of the next tile.

In hires bitmaps, screen RAM is used to hold the foreground and background colors of each tile (high and low nibble of each byte). This is the only VIC-II mode that does not make any use of the color RAM at $D800 or the background color register at $D021.

Multicolor bitmap mode allows three colors per tile (the fourth is the background color as set in $D021). Colors 1 and 2 are selected by the bits in screen RAM (same as hires bitmaps) and the third is from color RAM.

Despite the high level of color detail and all-points-addressable capabilities of bitmap mode, it is generally impractical for in-game graphics due to requiring a high amount of system resources (8k for the pixel data plus considerable more CPU cycles to modify each tile) and normally cannot be scrolled. Thus, it is normally only seen on loader and sometimes title screens. Dig Dug and Donkey Kong (Atarisoft) are two of the more notable examples of C64 games which utilize bitmap graphics.

Sprites

VIC-II sprites are either 24x21 monochrome or 12x21 multicolor. Similar to character graphics, the latter have one individual color for each sprite and two global ones. VIC-II has eight sprites, each of which uses 64 bytes of memory to store but, with certain limitations, in can display many more. Sprite multiplexing is a common method of getting more than eight on screen (although there still is a maximum of eight per scan line). The VIC-II scanline counters are polled until the desired point is reached on screen, after which the program quickly changes the sprite coordinates. This programming trick and other workarounds could result to over twenty sprites onscreen once. For a demo, though, the limit is considerably more flexible.

In theory the maximum number of different sprites visible at the same time is 256 (assuming the VIC-II's entire 16k page was filled). They are addressed by using a block number to refer to each sprite pattern in memory beginning with 0 and going to 255 ($FF) depending on their position in the video page. (if Page 2 is used, Block 0 would refer to the sprite stored at $4000 and Block 255 would be at $7FBF).

Each sprite may be double-sized vertically, horizontally or both. This does not make the sprite bigger (except visually) or add more pixels to the sprite, but merely upscales the existing pixels.

Because the horizontal position register for sprites is one byte and limited to a maximum value of 255, it could not cover the entire 320 pixels of the VIC-II's screen area, thus an additional register called the Most Significant Byte Flag is provided for this.

$D01E and $D01F contain the Background and Sprite-to-Sprite Collision registers. The former is rarely used because it cannot provide information on the specific background object the sprite is touching.

$D01B contains the Sprite To Background priority register, which is used to govern whether a sprite moves behind or in front of background objects. When a sprite enters the same space as another sprite, the lower-numbered ones will always pass over the higher numbered ones.

Scrolling

In order to scroll a character screen, the VIC-II is set to 38-column and/or 24-line mode via the registers at $D011 and $D016. This creates an off-screen buffer where the row of characters to be scrolled is placed. By adjusting the scroll bits in the above-mentioned registers, one row may be moved on-screen after which it repeats unless a new row is put in the buffer. Color RAM is scrolled simultaneous with screen RAM and works the same way.

VIC-II scrolling is a relatively complicated, CPU intensive task, although it was not uncommon for C64 game programmers to cheat by designing graphics so that the color RAM can remain static. Another standard trick is to cover the bottom or top 25% of the screen with a score counter to reduce the amount of scrolling that has to be performed. Finally, it is usually necessary to use an extra 1k piece of RAM to write character data to and then "blit" it into the screen RAM to prevent screen tearing, although this cannot be done with color RAM.

Raster interrupts

Utilization of raster interrupts is an essential part of C64 game programming. In the computer's power-on default state, the first CIA chip generates an interrupt 60 times per second which sends the CPU to the kernel IRQ handler at $EA31. This acknowledges the CIA interrupt, updates the clock, scans the keyboard, and blinks the cursor in BASIC. Games normally disable this and instead set up the VIC-II to generate interrupts when a specific scanline is reached, which is necessary for split-screen scrolling and playing music. The game remaps the IRQ vector at $0314/$0315 to its raster handler which performs these functions and then optionally executes a JMP $EA31 instruction to return control to the kernel, provided that the game does not take over the machine and therefore doesn't want to apply kernel's interrupt routine.

Some games use only one IRQ; however, nested ones are more common and improve program stability. In this setup, the IRQ is remapped to the second routine and so forth for each one until the last one restores it to the address of the first IRQ. When nested IRQs are used, only one JMP $EA31 instruction is needed in the chain and the others can be ended with JMP $EA81, which simply goes to the end of the kernel handler.

The VIC-II may also generate a raster interrupt from the collision registers, but this feature is rarely used.

Memory mapping

The VIC-II has a 14-bit address bus and can use any of the four 16k segments of the C64's memory space for video data. To manage this, two additional address bits are contributed by port bits of CIA. $0000-$3FFF is the power-on default. The second segment ($4000–$7FFF) is typically the best choice for games as it is the only segment that is completely free RAM with no ROMs or I/O registers mapped into it. The fourth segment ($C000–$FFFF) is also a good choice, provided that machine language is used, as the kernel ROMs must be disabled to gain read access by the CPU. Note that graphics data may be freely stored underneath the BASIC ROM at $A000-$BFFF, the kernel ROM at $E000-$FFFF or I/O registers at $D000–$DFFF, since the VIC-II only sees RAM, regardless how the CPU memory mapping is adjusted; character ROM is visible only in the first and third segment. The screen RAM, bitmap page, sprites, and character sets must all occupy the same segment window (provided the CIA bits aren't changed by scanline interrupt).

Registers

The VIC-II has 47 read/write registers listed below:

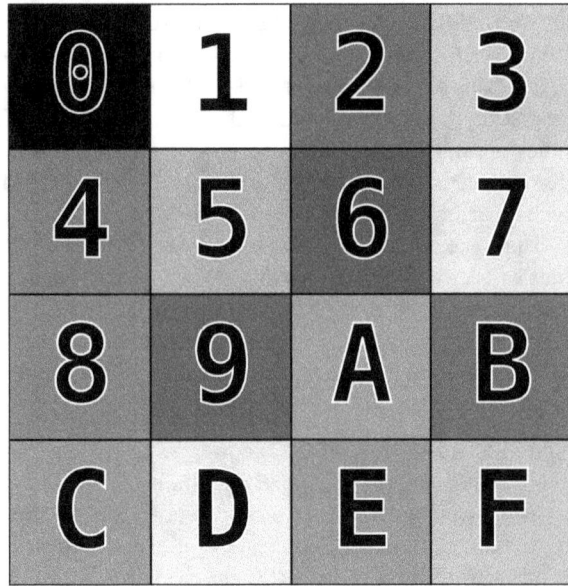

The VIC-II chip has a fixed 16-color palette, shown above.

Colors

In multicolor character mode (160×200 pixels, which most games used) characters had 4×8 pixels (the characters were still approximately square since the pixels were double width) and 4 colors out of 16 colors. The 4th color was the same for the entire screen (the background color), while the other 3 could be set individually for every such 4×8 pixel area. Two colors were loaded from the active text screen, and the third was loaded from color RAM. Sprites in multicolor mode (12×21 pixels) had three colors: two shared among all sprites and one individual. The artist had to pick shared colors such that the combination with individual colors led to a colorful impression. Some games reloaded shared colors during the raster interrupt; for example, the game *Turrican II's* underwater area (which was vertically distinct) had different colors. Others, such as Epyx's *Summer Games* and *COMPUTE!'s Gazette's Basketball Sam & Ed*, overlaid two high-resolution sprites to allow two foreground colors to be used without sacrificing horizontal resolution . Of course, this technique reduced the number of available sprites by half.

On PAL C64s, the PAL delay line in the monitor or TV which averages the color hue, but not the brightness, of consecutive screen lines can be used to create seven nonstandard colors by alternating screen lines showing two colors of identical brightness. There are seven such pairs of colors in the VIC chip.

The C64's team did not spend much time on mathematically computing the 16 color palette. Robert Yannes, who was involved with the development of the VIC-II, said:

I'm afraid that not nearly as much effort went into the color selection as you think. Since we had total control over hue, saturation and luminance, we picked colors that we liked. In order to save space on the chip, though, many of the colors were simply the opposite side of the color wheel from ones that we picked. This allowed us to reuse the existing resistor values, rather than having a completely unique set for each color.[7]

The VIC-IIe

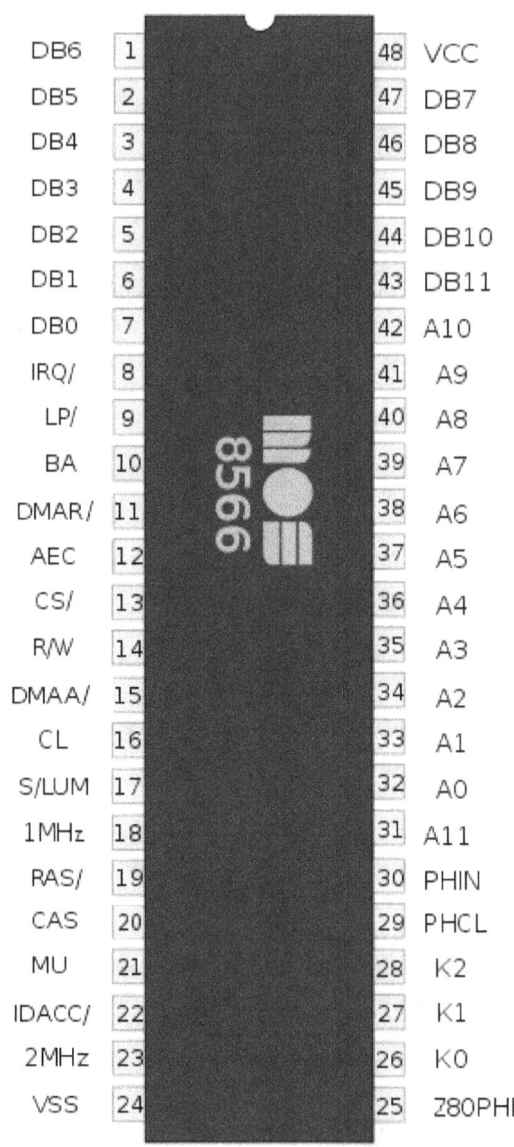

MOS 8566 VIC-IIe pinout

The 8564/8566 VIC-IIe in the Commodore 128 used 48 pins rather than 40, as it produced more signals, among them the clock for the additional Zilog Z80 CPU of that computer. It also had two extra registers. One of the additional registers was for accessing the added numerical keypad and other extra keys of that computer; this function was added to the VIC merely because that proved to be the easiest place in the computer to add the necessary three extra output pins. The other extra register was for toggling between a 1 MHz and a 2 MHz system clock; at the higher speed the VIC-II's video output is merely displaying every second byte in the code as black hires bit-pattern on the screen, suggesting use of the C128's 80-column mode at that speed (via the 8563 VDC RGB chip). Rather unofficially, the two extra registers were also available in the C128's C64 mode, permitting some use of the extra keys, as well as double-speed-no-video execution of CPU-bound code (such as intensive numerical calculations) in self-made C64 programs.[8] The extra registers were also one source of minor incompatibility between the C128's C64 mode and a real C64 - a few older C64 programs inadvertently wrote into the 2 MHz toggle bit, which would do nothing at all on a real C64, but would result in a messed-up display on a C128 in C64 mode.

The VIC-IIe has the little-known ability to create an additional set of colors by manipulating the registers in a specific way that puts the color signal out of phase with what other parts of the chip consider it to be in.

Using the specific behavior of the VIC-IIe's test bit, it is furthermore capable of producing a real interlace picture with a resolution of 320×400 (hires mode) and 160×400 (multicolor mode).

2.44.4   List of VIC-II versions

- PAL

  - MOS Technology 6569 – (PAL-B, used in most PAL countries)

  - MOS Technology 6572 – (PAL-N, used in southern South America only)

  - MOS Technology 6573 – (PAL-M, used in Brazil only)

  - MOS Technology 8565 – HMOS-II version for "C64E" motherboards

  - MOS Technology 8566 – VIC-II E (PAL-B) C128 version

  - MOS Technology 8569 – VIC-II E (PAL-N) C128 version

- NTSC

  - MOS Technology 6566 – designed for SRAM/non-muxed address lines (used in the Commodore MAX Machine)

- MOS Technology 6567 – Original NMOS version

- MOS Technology 8562 – HMOS-II version

- MOS Technology 8564 – VIC-II E C128 version

Notes

In all C64 models VIC-II is socketed for easy replacement, but it is important to notice that 6569, 6572, 6573, 6566 and 6567 use 12 volts and 5 volts when 8565 and 8562 use only 5 volts. Replacing old version with new version without motherboard modification destroys 8565 and 8562 if powered up in the oldest versions of C64 motherboards.

Several revisions of 6569 exist: 6569R1 (usually gold plated), 6569R3 and 6569R5. The most common version of 8565 is 8565R2.

2.44.5   See also

- Video Display Controller

- List of home computers by video hardware

2.44.6   References

[1] Bagnall, Brian (2005). "The Secret Project 1981". *On the Edge: The Spectacular Rise and Fall of Commodore* (1 ed.). Winnipeg, Manitoba: Variant Press. pp. 224–225. ISBN 0-9738649-0-7.

[2] Perry, Tekla S.; Wallich, Paul (March 1985). "Design case history: the Commodore 64" (PDF). *IEEE Spectrum* (New York, New York: Institute of Electrical and Electronics Engineers): 48–58. ISSN 0018-9235. Retrieved 2011-11-12.

[3] Bagnall, Brian (2005). "The Secret Project 1981". *On the Edge: The Spectacular Rise and Fall of Commodore* (1 ed.). Winnipeg, Manitoba: Variant Press. p. 227. ISBN 0-9738649-0-7.

[4] Bagnall, Brian (2005). "The Secret Project 1981". *On the Edge: The Spectacular Rise and Fall of Commodore* (1 ed.). Winnipeg, Manitoba: Variant Press. p. 229. ISBN 0-9738649-0-7.

[5] Bagnall, Brian (2005). "The Secret Project 1981". *On the Edge: The Spectacular Rise and Fall of Commodore* (1 ed.). Winnipeg, Manitoba: Variant Press. p. 230. ISBN 0-9738649-0-7.

[6] Bagnall, Brian (2005). "The Secret Project 1981". *On the Edge: The Spectacular Rise and Fall of Commodore* (1 ed.). Winnipeg, Manitoba: Variant Press. p. 242. ISBN 0-9738649-0-7.

[7] Timmermann, Philip. "Commodore VIC-II Color Analysis (Preview)". Retrieved 14 February 2015.

[8] Cowper, Ottis R.; Florance, David; Heimarck, Todd D.; Krause, John; Miller, George W.; Mykytyn, Kevin; Nelson, Philip I.; Victor, Tim (October 1985). "Chapter 7. System Architecture". *COMPUTE!'s 128 Programmer's Guide.* Greensboro, North Carolina: COMPUTE! Publications. pp. 348–349. ISBN 0-87455-031-9.

- "Appendix N: 6566/6567 (VIC-II) Chip Specifications". *Commodore 64 Programmer's Reference Guide* (PDF) (1 ed.). Commodore Business Machines. 1982. pp. 436–456. ISBN 0-672-22056-3.

2.44.7 External links

- The MOS 6567/6569 video controller (VIC-II) and its application in the Commodore 64 - detailed hardware description of the VIC-II.

- Commodore VIC-II Color Analysis (Preview) - an attempt to provide accurate information as to the VIC-II color palette, by Philip Timmermann.

- Description of C64 graphics modes - simple explanations with example pictures of the common modes used for C64 graphics, including hacked and software-assisted modes.

- Real Interlace video modes using the VIC-IIe.

- VIC programming information on Codebase64.

- VIC-II die shots

2.45 PETSCII

PETSCII (*PET Standard Code of Information Interchange*), also known as **CBM ASCII**, is the character set used in Commodore Business Machines (CBM)'s 8-bit home computers, starting with the PET from 1977 and including the VIC-20, C64, CBM-II, Plus/4, C16, C116 and C128.

2.45.1 History

The character set was largely designed by Leonard Tramiel (the son of Commodore CEO Jack Tramiel) and PET designer Chuck Peddle. The graphic characters of PETSCII were one of the extensions Commodore specified for Commodore BASIC when laying out desired changes to Microsoft's existing 6502 BASIC to Microsoft's Ric Weiland in 1977.[1] The VIC-20 used the same pixel-for-pixel font as the PET, although the characters appeared wider due to the VIC's 22-column screen. The Commodore 64, however, used a slightly re-designed, heavy upper-case font, essentially a thicker version of the PET's, in order to avoid color artifacts created by the machine's higher resolution screen. The C64's lowercase characters are identical to the lowercase characters in the Atari 800's system font (released several years earlier).

Peddle claims the inclusion of card suit symbols was spurred by the demand that it should be easy to write card games on the PET (as part of the specification list he received).

2.45.2 Specifications

C64 startup screen with shifted and unshifted modes of PETSCII, and the two characters from ASCII-1963.

PETSCII is based on the 1963 version of ASCII (rather than the 1967 version, which most if not all other computer character sets based on ASCII use). Assuming the graphics mode is **unshifted**, PETSCII has only uppercase letters in its powerup state, an up-arrow (↑) instead of a caret (^) in position $5E and a left-arrow (←) instead of an underscore (_) in position $5F. Also, in the VIC-20 and C64 version, the backslash (\ ) in position $5C is occupied by a British pound sign (£). In *unshifted mode*, codes $60–$7F and $A0–$FF are allotted to CBM-specific block

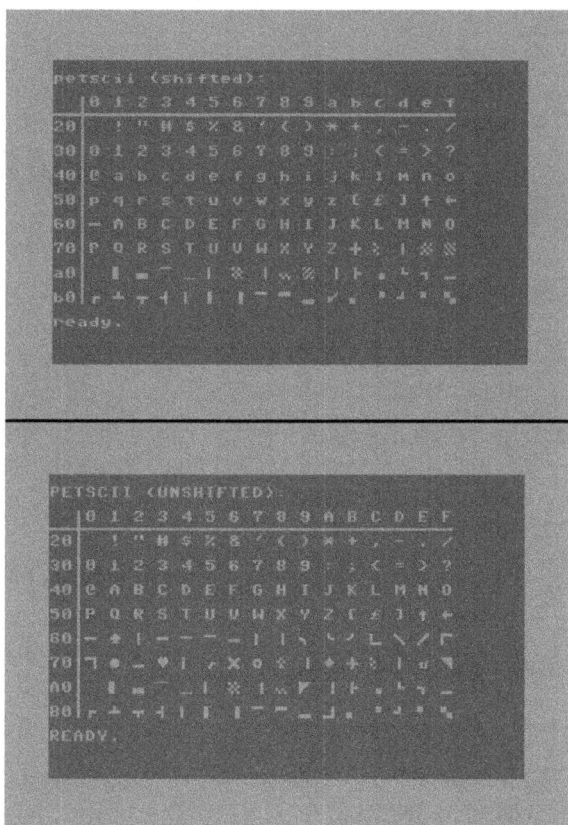

PETSCII Chart as displayed on the C64 in shifted and unshifted modes. (Not shown are control codes, as well as characters in the $C0-$FF range, which are the standard uppercase keycodes returned from the keyboard, and which are mirrored to the range $60-7F)

graphics characters (horizontal and vertical lines, hatches, shades, triangles, circles and card suits). Ranges $00–$1F and $80–$9F have control codes.

The Commodore PET's lack of a programmable bitmap-mode for computer graphics, as well as it having no re-definable character set capability, may be one of the reasons PETSCII was developed; by creatively using the well thought-out block graphics, a higher degree of sophistication in screen graphics is attainable than by using plain ASCII's letter/digit/punctuation characters. In addition to the relatively diverse set of geometrical shapes that can thus be produced, PETSCII allows for several grayscale levels by its provision of differently hatched checkerboard squares/half-squares. Finally, the reverse-video mode (see below) is used to complete the range of graphics characters, in that it provides mirrored half-square blocks.

PETSCII also has a *text mode*, in which lowercase letters occupy the range $41–$5A, and uppercase letters occupy the range $C1–$DA. The text mode is not available at powerup, but must be actuated by holding one of the SHIFT

keys and then press and release the *Commodore* key. Regardless of whether the chip has undergone this graphic "shift", there are block graphic characters in the range of $E0-FF. This serves to distinguish PETSCII from those kinds of ASCII that go back no farther than ASCII-1967, so any text transfer between an 8-bit Commodore machine and one that uses 1967-derived ASCII would result in text where uppercase letters appear to be lowercase, and lowercase letters uppercase. There is no easy Boolean operation to change these cases to the proper case. Thus, as with other computers based on non-standard-ASCII character sets, software conversion is needed when exchanging text files and/or telecommunicating with standard ASCII systems. The other ranges are unchanged in shifted mode; this means that the other characters added in ASCII-1967 besides lowercase letters — i.e. the grave accent, curly braces, vertical bar, and tilde — do not exist in PETSCII.

Included in PETSCII are cursor and screen control codes, such as {HOME}, {CLR}, {RVS ON}, and {RVS OFF} (the latter two activating/deactivating reverse-video character display). The control codes appeared in program listings as reverse-video graphic characters, although some computer magazines, in their efforts to provide more clearly readable listings, pretty-printed the codes using their actual names, like the above examples. Such names were commonly enclosed in curly braces in the listings. This prevented ambiguity, since, as mentioned, PETSCII had no curly brace characters. The screen control codes were essentially similar to escape codes for text based computer terminals.

As indicated above, PETSCII provides for shifting between the power-on default (unshifted) uppercase+graphics character set and the alternative (shifted) lower+uppercase set (where the shifted set contains a subset of the block graphic characters of the unshifted set). The shift between modes is done by POKEing location 59468 with the value 14 to select the alternative set or 12 to revert to standard. On C64 the sets are alternated by flipping bit 2 of the byte 53272. On some models of PET this can also be achieved via special control code PRINT CHR$(14) which adjust the line spacing as well as changing the character set; the POKE method is still available and does not alter the line spacing.[2] Thus, screen editor state changes, rather than the employment of separate ASCII codes, are used to choose between single-case (all capitals) and dual case. In the VIC-20, C64, and later machines (not including the CBM business computers), color codes supplement the other screen control codes. (The colors of the VIC-20 and C64/128 are listed in the VIC-II article.)

2.45.3 Codepage layout

Since not all of the characters encoded by PETSCII are 'graphic' (i.e., control codes) and not all of them have a corresponding Unicode representation, they cannot be portably displayed in a web browser. The following table shows the glyphs for PETSCII graphic characters where there is a corresponding Unicode glyph, and the Unicode replacement character U+FFFD (�) otherwise. Control characters and other non-printing characters are represented by abbreviations for their names. Where a particular code point encodes both a shifted and unshifted character, both characters are shown, with the unshifted character on the left. Row and column headings indicate the hexadecimal digit combinations to produce the eight-bit code value; e.g., the letter *L* is at code value 4C.

Note that the table below is for the Commodore 64. Other Commodore machines used slightly different versions of PETSCII, which used different control characters and in some cases different graphic characters. For example, on the Commodore 128 $07 was the bell control character, and on CBM machines prior to the VIC-20, characters $2C and $6C both produced a comma character, albeit with slightly different semantics.[3]

The actual character generator ROM used a different set of assignments. For example, to display the characters "@ABC" on screen by directly POKEing the screen memory, one would POKE the decimal values 0, 1, 2, and 3 rather than 64, 65, 66, and 67.

Some PETSCII Codes can't be printed and are only used for Keyboard input (e.g. F1, RUN/STOP).

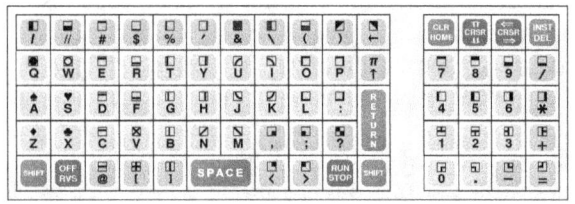

PET 2001 keyboard layout, illustrating PETSCII graphics characters

2.45.4 See also

- ATASCII

- ZX Spectrum character set

- Extended ASCII

- Text semigraphics

2.45.5 Notes

1. ^ The Amiga home/personal computer family uses standard ISO-8859-1.

2. ^ see *On The Edge* by Brian Bagnall, ISBN 0-9738649-0-7, page 43, 54-55.

2.45.6 References

[1] A Conversation with Chuck Peddle, Bil Herd, Jeri Ellsworth - part 3 (2009 videoconference, 06:30)

[2] THE COMMODORE PET COMPUTER / FREQUENTLY ASKED QUESTIONS FILE - VERSION 1.7 (Updated 25 November 2000) BY LARRY ANDERSSON, COMMODORE COLLECTOR AND PET ENTHUSIAST

[3] Commodore Trivia Edition #26 Answers for February 1996

2.45.7 External links

- PETSCII character map, part 1, part 2, part 3 (JPEG)

- An attempt at PETSCII to Unicode mapping, unshifted, shifted

- Commodore 128 PETSCII control characters

- Typography in 8 bits: System fonts

- Online PETSCII-art editor

2.46 PlayNET

PlayNet (or **PlayNET**) was a U.S. online service for Commodore 64 personal computers that operated from 1984 to 1987. It was operated by the PlayNet, Inc of Troy, New York.

2.46.1 History

PlayNet was founded [1] in 1983 by two former GE Global Research employees, Dave Panzl and Howard Goldberg,[2] as the first person-to-person, online communication and game network to feature home computer based graphics. The PlayNet software became the foundation for the first version of what is today America Online.

The founders launched the business initially with their own money. They then raised over $2.5 million from a variety of investors, including the venture capital funds of the Town of North Greenbush NY, Key Bank, Alan Patricof & Associates, and the New York State Science and Technology

Foundation, and a group of individual investors through a limited R&D partnership led by McGinn Smith.

In 1985 PlayNet licensed [3] their system to Control Video Corporation (CVC, later renamed Quantum Computer Services), which in October 1991 changed its name to America Online. The modified version of the PlayNet software (Quantum Link or **Q-Link**) was ported by Quantum to the PC to create the first version of the AOL software. As recently as 2005, some aspects of the original PlayNet communication protocols still appeared to be used by AOL.

The PlayNet offices were initially located in the J Building on Peoples Avenue in Troy, NY part of the Rensselaer Polytechnic Institute incubator program. It subsequently moved to RPI's Technology Park in North Greenbush NY. At its peak, PlayNet employed 30 people including software developers, customer service staff, etc.

The maximum number of subscribers was approximately 3000, with up to around 200 logged in at a time. PlayNet declared bankruptcy in February, 1986 and ceased operations in 1988 after Quantum stopped paying royalties.

The service had two membership options, an $8/month service charge plus $2.75/hour connect time charge, or no service charge and $3.75 per hour connection charge. File downloads were charged a flat rate of $0.50 each

2.46.2 Software details

PlayNet was originally designed around online interactive games which allowed chatting while playing. PlayNet also featured electronic mail, online chat, bulletin boards, file sharing libraries, online shopping, and instant messaging (using *On Line Messages*, or *OLM*s). Games were mostly 'traditional' games and some well-known boardgames. Games were programmed in a mixture of BASIC and assembly language.

Unlike other online systems of the era, PlayNet was highly graphical and required client software, and included error correction in the communication protocols. The server software for PlayNet ran on Stratus fault-tolerant computers and was written in PL/1. AOL continued to use Stratus computers and parts of the PlayNet server software until the late 1990s or later.

- The client software on the Commodore 64 ran a multitasking pseudo-operating system based on a Finite State Machine language.

2.46.3 Game list

- Checkers

- Chess

- Backgammon

- Hangman

- Bridge

- Stratego

- Connect 4

- Chinese Chess

- Go

- Several others

Games/features never finished/released:

- Multiplayer Dungeons and Dragons

- Poker

- Various other card games and wargames

- Auditorums and panel discussions

The never-released D&D game was very similar to Neverwinter Nights (AOL game).

Connections to PlayNet were made by modems at 300 baud via X.25 providers such as Tymnet and Telenet. In 1985, pricing was $6 per month, with additional fees of $2 per hour, after a one-time membership fee of $30.[4]

The system competed with many other online services like CompuServe and The Source (service), as well as Bulletin board systems (single or multiuser). PlayNet's graphical display was better than many of these competing systems because it used specialized client software with a nonstandard protocol. However, this specialized software and nonstandard protocol limited its market to the Commodore 64.

In 2005, hobbyists managed to reverse engineer the communications protocol and allow people running the QuantumLink software on an emulator or original hardware (via a serial cable) to run a reduced version of the service called Quantum Link Reloaded.

2.46.4 Reception

Ahoy! in 1986 called PlayNET "one of the best values around for Commodore users". The reviewer stated that he had found the network's users "to be just about the friendliest group of people around", but criticized the slow disk load times and the network's weekday hours of operation.[5]

2.46.5 References

[1] "On the Way to the Web: The Secret History of the Internet and Its Founders", by Michael A. Banks, 2008, pages 91, 189.

[2] Early news story about PlayNET: http://www.youtube.com/watch?v=1W_yfq1CLwY

[3] AOL history: http://en.wikipedia.org/wiki/AOL

[4] Lockwood, Russ. "Tracking the Affordable Home Service." *A+: The Independent Guide for Apple Computing*, April 1985: 54-56.

[5] Behling, B. W. (1986-01). "PlayNET". *Ahoy!*. pp. 81–84. Retrieved 3 July 2014. Check date values in: |date= (help)

2.46.6 External links

- PlayNET page

- Remember Q-Link

- Remembering Q-Link

- AOL Disk Collection: Q-Link

- Quantum Link Reloaded

- RUN Magazine Issue 20

2.47 Punter (protocol)

Steve Punter in BBS: The Documentary.

Punter is a protocol for file transfer developed in the 1980s by Steve Punter. There are various types of Punter such as PET Transfer Protocol (PTP), C1 and C2.

2.47.1 PET Transfer Protocol

The **PET Transfer Protocol (PTP)**, also known as **Punter** or **Old Punter**, was developed ca. 1980 by Steve Punter for use with his PETBBS and BBS64 bulletin board system (BBS) software. The "PET" in the name comes from the Commodore PET computer.

Compared to other contemporary protocols, PTP is slower than YMODEM and ZMODEM but faster and more reliable than XMODEM.

The earliest version of Punter supports only 7-bit transfers and uses a back-correction algorithm involving two checksums for failsafes. One of the two checksums is additive, and the other is Boolean in nature (executing EOR instructions), making for an easy to understand algorithm for other programmers to understand and emulate. Having two checksums--both of them being 16 bits wide--makes it significantly more accurate than the single-byte checksum used by XMODEM, its major competitor in the early 1980s. Regardless of the potential for errors to creep in, in comparison to the YMODEM protocol of the late 1980s, which is arguably superior, it has been widely used on Commodore PET and Commodore 64 based bulletin boards.

2.47.2 What It Looks Like to the User

Not all of the transmission is visible to the user. The most noticeable part of the transmission is the report of status codes like ACK, GOO, BAD, and SYN for handshaking results. A typical transmission might look like "ACKGOOGOOGOOGOOBADGOO-GOOGOOBADGOO," with bad blocks reported to the user just as frequently as they occurred. This allows users to record the error rate according to hour and day of the week, and determine which hours of the day and which days of the week had cleaner phone lines. Unlike modern computers, the C64 and C128 can poll the User Port (where the modem is interfaced) at slightly different BAUD rates and connection speeds. For instance, a transmission at 1200 BAUD on Sunday evening might actually produce fewer errors than 2400 BAUD on Tuesday afternoon. By choosing slower BAUD rates, files can actually be transmitted faster than at higher BAUD rates, inasmuch as there are fewer resendings in a given transfer.

2.47.3 C1

C1, also known as **New Punter**, was developed in 1984 by Steve Punter as a successor to PTP. C1 was the standard protocol for use on Commodore BBSes, and was rarely supported by terminal or BBS software for other operating systems.

The C1 specification was rife with inaccuracies and ambiguities, making it difficult to implement from scratch. Nevertheless, the protocol came into widespread use because Punter released the source code for the original implementation into the public domain.

Technical information

C1 could transmit block sizes up to 255 bytes with a recommended (but not enforced) minimum of 40 bytes and an overhead of 7 bytes per block. It is optimized for transferring files stored on Commodore computers, whose DOS treats executable, sequential, and random-access files identically.

2.47.4 Multi-Punter

The term **Multi-Punter** can refer to any one of three or four mutually incompatible third-party variants of C1 which permit batch-file transfers, as opposed to C1, which was designed for single-file transfers.

One such variant, C2, also known simply as **Punter**, was developed circa 1985 by Steve Punter. Like C1, it is optimized for transferring files stored by 8-bit Commodore computers.

Another variant was developed circa 1987 by Alan Peters.

2.47.5 External links

- Steve Punter's original C1 specification

- C1 specification with interpretive annotations

- Punter program for IBM PC systems

- Notes on implementing C1 and Peters' Multi-Punter

2.48 Quantum Link

"Q-Link" redirects here. For the television technology, see AV.link.

Quantum Link (or **Q-Link**) was a U.S. and Canadian online service for Commodore 64 and 128 personal computers that operated from November 5, 1985, to November 1, 1995. It was operated by Quantum Computer Services of Vienna, Virginia. In October 1991 they changed the name to America Online, which continues to operate the AOL service for the IBM PC and Apple Macintosh today. Q-Link was a modified version of the PlayNET system, which

Control Video Corporation (CVC, later renamed Quantum Computer Services) licensed.

Q-Link featured electronic mail, online chat (in its People Connection department), public domain file sharing libraries, online news, and instant messaging (using *On Line Messages*, or *OLM*s). Other noteworthy features included online multiplayer games like checkers, chess, backgammon, hangman and a clone of the television game show "Wheel Of Fortune" called 'Puzzler'; and an interactive graphic resort island called Habitat while in beta-testing and later renamed to Club Caribe.

In October 1986 QuantumLink expanded their services to include casino games such as bingo, slot machines, blackjack and poker in RabbitJack's Casino and RockLink, a section about rock music. The software archives were also organized into hierarchal folders and were expanded at this time.[1]

In November 1986 the service began offering to digitize users' photos to be included in their profiles, and also started an online auction service.[2]

Club Caribe was developed with Lucasfilm Games. It was designed using software that later formed the basis of Lucasfilm's Maniac Mansion story system (SCUMM). Users controlled on-screen avatars that could chat with other users, carry and use objects and money (called *tokens*), and travel around the island one screen at a time. One fun note - Club Caribe allowed you to take the head off of your character, carry it around or even set it down. However, if you did set it down, then someone else could pick it up and carry it away. Hence the reason for some headless people walking around the island. Club Caribe was a predecessor to today's MMOGs.

Connections to Q-Link were typically made by dial-up modems with speeds ranging from 300 to 2400 baud, with 1200 baud being the most common. The service was normally open weekday evenings and all day on weekends. Pricing was $9.95 per month, with additional fees of six cents per minute (later raised to eight) for so-called "plus" areas, which included most of the aforementioned services. Users were given one free hour of "plus" usage per month. Hosts of forums and trivia games could also earn additional free plus time.

Q-Link competed with other online services like CompuServe and The Source, as well as bulletin board systems (single or multiuser), including gaming systems such as *Scepter of Goth* and *Swords of Chaos*. Quantum Link's graphic display was better than many competing systems because they used specialized client software with a nonstandard protocol. However, this specialized software and nonstandard protocol also limited their market, because only the Commodore 64 and 128 could run the software necessary to access Quantum Link.

In the summer of 2005 Commodore hobbyists reverse engineered the service, allowing them to create a Q-Link protocol compatible clone called **Quantum Link Reloaded** which runs via the Internet as opposed to telephone lines. Using the original Q-Link software as a D-64 file, it can be accessed using either the VICE Commodore 64 emulator (available on multiple platforms, including Windows and Linux), or by using authentic Commodore hardware connected to the Internet by way of a serial cable connected to a PC with internet access.

2.48.1 See also

- *Habitat*

- *Club Caribe*

2.48.2 References

[1] "RUN Magazine Issue 34".

[2] "RUN Magazine Issue 35".

2.48.3 External links

- Q-Link Promotion Video from 1986

- Quantum Link Reloaded

- Remember Q-Link

- JohnD39's Q-Link contact/memories site

- Keith Elkin's History of Q-Link

2.49 Scene World Magazine

Scene World Magazine (abbreviated *SWO*) is a disk magazine for the Commodore 64 home computer. The magazine has been released regularly since February 2001.

2.49.1 History

Scene World was founded in November 2000 by several Commodore scene personalities under the organization of Joerg "Nafcom" Droege. The initial magazine presentation system was programmed by Robin Harbron, who would later find success as one of the developers of the C64 Direct-to-TV game device.[1] Harbron stopped actively supporting the magazine in 2001; the presentation system has since

Scene World Magazine

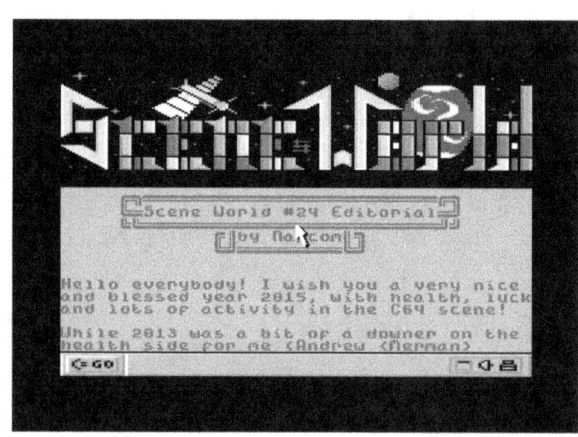

Screenshot of the presenter system of issue 24 of Scene World *Magazine displaying the editorial text*

been documented, modified, and updated by various editors and staff members.

Throughout *Scene World's* history, it has attempted to style itself as both an NTSC (North America, Japan, and South Korea) and PAL (Europe, Oceania, and the Middle East) production, allowing it to court talent and reach audiences in largely disparate computer cultures.[2] *Scene World* has also been one of the few disk magazines to actively seek individuals that do not fit into the specific software cracking or demo scene subcultures that most other disk magazines focus on. As a result, it has been able to conduct interviews with numerous non-Commodore-related computer industry pioneers.

Scene World has received media attention on several occasions. In March, 2001, Droege was interviewed by the German Radio channel Bayern 3[3] In October 2005, Droege was interviewed in the German magazine Lotek64[4] and in 2013 in German magazine LOAD #2.[5] In August 2015, Droege was recognized for his work on *Scene World* with a trading card (#2296) the Walter Day Collection [6]

Additionally, *Scene World* partnered with fellow disk magazine Loadstar (ISSN 0886-4144) in 2003 for a "Wild West" cross-promotion.[7]

2.49.2   Magazine Content

Disk sleeve of Scene World *Magazine issue 24 done by Indian designer Ishita Mukherjee*

Issues of the magazine are delivered as downloadable disk image files for use in emulators or on actual Commodore hardware. Text of the magazine is provided via presentation software. This software allows users to select and read articles and scroll through text, while also allowing them to change text and background color, switch between fonts, alternate between logos, and select either music or silence. Uniquely, support for the Commodore 1351 computer mouse is a feature of the presenter system;[8] during initial planning, Droege felt that American users would be more inclined to view the magazine if it resembled Berkeley Softworks' GEOS operating environment.

The text of the magazine is divided into articles or chapters, and covers topics ranging from news and updates, interviews (text-only or transcribed from video interviews), opinion and editorial content, demo party reports, and release charts.

In addition to Commodore-specific files, a PDF floppy disk sleeve image is also included with each issue. These "diskcovers" can be printed to create protective sleeves for people using physical floppy disks.

2.49.3   Further Endeavors

In 2012, Droege expanded *Scene World's* non-disk magazine activities to focus on video-based interviews, initially with Michael Tomczyk and John Draper, and later with other technology pioneers such as Martin Cooper, Jeroen Tel, Bil Herd, Chuck Peddle, Yash Terakura,[9] Walter Day,[10] James Bach, Alexey Pajitnov, Stewart Cheifet,[11] Chris Huelsbeck,[12] Jeri Ellsworth, and Ralph H. Baer. Baer's interview, in particular, garnered significant attention, it being the final—and longest—interview he gave prior to his death in December 2014.[13]

In July 2014, *Scene World* again expanded with an audio podcast, hosted by editorial staffer AJ Heller (with Droege as co-host), to focus primarily on technology personalities and newsmakers that are currently active and promoting products or services. Guests to date have included copyright, social network, security and privacy journalist Lars Sobiraj,[14] Frederik Schreiber and Mike Nielsen of 3D Realms,[15] Matt Falcus and Sven Vößing of Cinemaware,[16] SiREN and Esper from Ubisoft's Frag Dolls professional gaming team, and Charles Martinet, the voice actor for, among other things, Nintendo's Super Mario franchise,[17] Jay Maynard, cosplayer who made a costume based on the Tron movie becoming famous as the Tron Guy.[18]

In June 2015, Droege further expanded *Scene World's* online presence with a channel on the streaming platform Twitch.tv. This channel currently hosts live interviews, conducted by Droege,[19] and there are plans to utilize the channel to create original programming, as well as to stream the staff's participation in live and charity events.

Scene World currently utilizes an all-volunteer editorial staff of 21, and produces two issues of the disk magazine per year, generally in summer and winter, with video interviews scheduled to coincide with the disk magazine releases. The podcast, conversely, is released on a semi-monthly basis, and does not follow the release schedule of the magazine.

2.49.4   Charity and Fund Raising

Following the podcast with UbiSoft's Frag Dolls professional gaming team, *Scene World* staff members Droege and Heller committed to participating in the Extra-Life charity organization. The two are currently raising money for the Illuminate team (former Frag Dolls team).[20]

2.49.5   See also

Commodore 64

Disk magazine

List of disk magazines

2.49.6 References

[1] Patterson, Blake. "'P1XL Party' - A Multi-Game Pack for the Retro Lovers Out There". *http://toucharcade.com*. toucharcade. Retrieved 26 April 2015.

[2] Adok. "Diskmag Galore". *http://www.hugi.scene.org*. Hugi. Retrieved 25 April 2015.

[3] Kummert, Florian. "Der C64 lebt!". *http://www.br.de/radio/bayern3/*. Bayern 3 Radio. Retrieved 25 April 2015.

[4] Rust, Volker. "Scene World: PAL und NTSC kommen sich naeher". *http://www.lotek64.com*. Lotek64. Retrieved 25 March 2015.

[5] Riebe, Marco. "Das C64-Magazin Scene World". *http://load-magazin.de*. LOAD. Retrieved 25 April 2015.

[6] Todd. "The Collection". Walter Day. Retrieved 24 September 2015.

[7] Merman. "Loadstar: end of an era!". *http://lemon64.com/*. Lemon64 Forum. Retrieved 25 April 2015.

[8] Droege, Joerg. "Scene World presentation on HomeConnected meeting". *youtube*. TMCRole. Retrieved 25 April 2015.

[9] Kitty. "Scene World – Video interview with Yash Terakura". *http://www.vintageisthenewold.com/*. Vintage Is The New Old. Retrieved 25 April 2015.

[10] "News: Walter Day Skype Video Interview in Scene World Magazine". *http://www.twingalaxies.com*. Twin Galaxies. Retrieved 25 April 2015.

[11] Ditta, Sheraz. "Computer Chronicles fan site". *http://www.stquantum.com*. stquantum.com. Retrieved 25 April 2015.

[12] Kitty. "Scene World – Video interview with Chris Huelsbeck". *http://www.vintageisthenewold.com*. Vintage Is The New Old. Retrieved 25 April 2015.

[13] Freundorfer, Stephan. "Nachruf auf Ralph H. Baer - Der Vater des Videospiels". *http://www.gamestar.de*. GameStar. Retrieved 25 March 2015.

[14] "Zu Gast bei Scene World". *http://lars-sobiraj.de*. Retrieved 26 April 2015.

[15] "3D Realms Featured on the Latest Scene World Podcast". *http://www.3drealms.com*. Retrieved 27 March 2015.

[16] "Scene World Podcast Episode #2 The Return of Cinemaware". *http://www.commodorefree.com*. Commodore Free. Retrieved 25 April 2015.

[17] "Charles Martinet (Voice of Mario) Podcast interview". *http://www.playright.dk*. Retrieved 25 April 2015.

[18] Heller, AJ. "Podcast Episode #10 - Jay Maynard: The Tron Guy". *http://sceneworld.org*. Scene World Podcast. Retrieved 30 August 2015.

[19] "Scene World Twitch.TV Archive playlist". *www.youtube.com*. Retrieved 4 July 2015.

[20] "Extra Life. Play Games. Heal Kids.". *www.extra-life.org*. Retrieved 4 July 2015.

-
-
-
-

2.49.7 External links

- *Scene World* Magazine - Official Website
- *Scene World* on the Commodore 64 Scene Database
- *Scene World* on Pouet
- *Scene World* Podcast
- *Scene World* Podcast on iTunes
- *Scene World* Podcast torrent downloads via BitLove
- *Scene World* Video productions on YouTube
- *Scene World* live video podcasts on Twitch

2.50 Stack Light Rifle

The **Stack Light Rifle** is a light gun that was manufactured by Stack Computer Services and created for the ZX Spectrum, Commodore 64, and the Commodore VIC-20. It was released in 1983. The rifle is bundled with three games on tape, High Noon, Shooting Gallery and Grouse Shoot for the Spectrum. Different games were offered for the Commodore 64 and VIC-20 versions (all the games for these two systems were included on one cassette). It retailed for about $60, which is extraordinarily expensive given the fact that most cartridges were $10–20 each. The Stack Light Rifle is differentiated from future light guns as being very realistic looking; future unrealistic light guns such as the NES Zapper and the Sega Light Phaser dealt with controversy due to the guns still being misidentified as real firearms.

The main pistol is attached to 12 feet of cable which ends in a dead-ended ZX81-size connector which plugs into the Spectrum's user port. A barrel, stock and telescopic sight can all be attached to the pistol. The barrel actually facilitated the gun's performance as it filtered out ambient light. These three parts combined to provide a reasonable - if not perfect - degree of accuracy, and allowed the user to effectively use the light gun from the comfort of an armchair. One can extrapolate that the multi-part design was later mimicked on the Sega Menacer.

Variants of the Light Rifle were available for the ZX Spectrum, Commodore VIC-20 and Commodore 64 and all perform the same function. Like the Atari XG-1 light gun, the Stack Light Rifle was treated by the hardware as a light pen. Due to lack of availability of software drivers for the Light Rifle, only the three games that came with the device were available. In April 1985, Sinclair User magazine reported that Stack Computer Services company disappeared.

2.50.1 Technical specifications

The main component of the Stack Light Rifle System is the electronic target pistol that is connected to the computer by a generous length of lead. At the computer end, depending on the version, there is a connector for the appropriate socket or edge connector. On the ZX Spectrum version the connector contains two chips and a couple of simple components to interface the main electronics inside the gun to the computer. To make the pistol more accurate and to turn it into a rifle - it is supplied with a shoulder stock that clips and secures to the rear of the pistol, a barrel and a make-believe telescopic sight.

The electronics inside the pistol consist of a light detector or photo-diode and a small amplifier and buffer. Light coming down the barrel is focused by a small plastic lens onto the photo-diode, and the device is sensitive enough to detect the changes in intensity of the picture. Once boosted by the amplifier, the signal is clipped to provide a digital pulse rather than an analogue waveform and is then fed to the computer via the switch. The screen position that is being scanned at that moment is the position the rifle is pointing at. As the computer receives the pulse from the Light Rifle it compares the value of its scan registers with the screen position of the target and, if a match is found, the played has scored a direct hit.

2.50.2 Supported Games

Commodore 64

- Escape From Alcatraz
- High Noon[lower-alpha 1]
- Glorious 12th[lower-alpha 1]
- Gallery[lower-alpha 1]
- Crowshoot
- Rat's & Cats

VIC-20

- High Noon[lower-alpha 1]
- Glorious 12th[lower-alpha 1]
- Gallery[lower-alpha 1]

ZX Spectrum

- High Noon[lower-alpha 1]
- Invasion Force by Micromania

[1] bundled by Stack

2.50.3 External links

- Sinclair User Magazine: Issue 37, April 1985
- World of Spectrum Feature on the Stack Light Rifle
- *Stack Light Rifle* at World of Spectrum

2.51 Super 1750 Clone

The **Super 1750 Clone** was a 512 kB RAM expansion unit designed as a tiny, but compatible, third-party replacement for Commodore's then out-of-production CBM 1750 REU. Manufactured by Chip Level Designs, the Super 1750 Clone was sold by Software Support International.

- Used the same MOS 8726 REC (RAM Expansion Controller) chip as the Commodore REUs.
- Worked on the C128 and the C64.
- Rather than 16 chips of 256K×1-bit DRAMs, it used four 256K×4-bit DRAMs (in ZIP packages). This gave several advantages over Commodore's original REUs:
  - Less power consumption, so did not require the C64 to be upgraded with a heavy-duty power supply to use it.
  - Much smaller; the plastic case was the same type used by the Epyx *FastLoad* cartridge.

2.51.1 External links

- c64.org: Super 1750 Clone User's Guide (text)
- c64.org: Super 1750 Clone User's Guide (zip file) (eText)

2.52 Super Expander 64

The **Super Expander 64** was a cartridge-based extension to the built in BASIC V2 interpreter of the then immensely popular Commodore 64 home computer: Since the 64 was developed in a hurry, Commodore simply adapted the BASIC V2 from the PET line of computers and the VIC 20 for their new machine, with no support for the advanced sound and graphics capabilities of the 64. To make use of the advanced hardware, BASIC programmers needed to memorize hardware addresses and "POKE" commands directly to the memory-mapped devices. Later 8-bit systems from Commodore had BASIC interpreters enhanced to support the special hardware, and with the Super Expander 64 cartridge, Commodore 64 users could "retrofit" their machine with a BASIC to match the hardware capabilities.

The extra code was mapped into the "lower cartridge" 8 kilobytes area at $8000-$9FFF, thus reducing the 38911 bytes for user programs by said 8K.

2.52.1 Graphics

After initializing the screen for "high-resolution" (320 × 200 monochrome pixels) or multicolor (160 wide × 200 pixels in four colors) graphics with the GRAPHIC command, one could draw lines, circles, ellipses, arcs, boxes and more using the DRAW, CIRCLE, and BOX commands. PAINT would "flood-fill" an area enclosed by lines, e.g. the interior of a CIRCLE or BOX. A CHAR command was used to "print" characters from the character generator ROM onto the bitmap graphics screen. SSHAPE and GSHAPE would store the contents of a rectangular area of the high-res graphics into a string variable, and GSHAPE would "stamp" it back onto the screen at arbitrary locations. Such "graphics-in-a-string" could also be used to transfer something drawn on the hi-res screen into one of the eight sprite patterns.

Sprites

Besides a range of commands to initialize, position and move sprites (or Movable Object Blocks as Commodore called them; hardware-supported graphic elements that could move freely on the screen independently of other graphics and text on the screen), Super Expander had a built in tool to edit the pattern of 8 sprites (called upon with the SPRDEF command), either in high-res (24 × 21 pixels) or multicolor (12 wide × 21 pixels) mode.

There was even a way of implementing "interrupts" in the BASIC program if two sprites collided, if a sprite collided with other graphics and/or text on the screen, or if an attached light pen was activated. A COLINT command set up the interrupt, pointing to the beginning BASIC line number of the "interrupt handler", which had to end in a RETURN statement (part of standard, unexpanded BASIC) in order to transfer control back to the interrupted, "mainline" part of the program.

2.52.2 Sound

Playing a sequence of musical notes was hooked onto the standard BASIC "PRINT" command by the use of a special "control character", much like the cursor control, color changes and other control characters. E.g. PRINT CHR$(6);"CDEFGAB" played a rising scale. Commands like TEMPO and TUNE was used to set the playback tempo and the timbre of the note sequence.

A "quirk" of this feature is that by typing the special control character (by pressing CTRL + F) along with a quote mark, then deleting the quote mark, the machine would "play" whatever was typed while editing the program; hit the G key, and the machine played a "Pling!" with the pitch of a G note...!

2.52.3 Hardware I/O

Functions like RJOY, RPEN and RPOT would read the state of a connected joystick, light pen, or analog "paddle". In combination with the sprite-motion-related commands, it only took a single line of BASIC code to make a sprite move in the direction indicated by a connected joystick.

2.52.4 Miscellaneous

A KEY command was available, which would set up the four function keys on the 64's keyboard to "enter" an arbitrary string. By default, these keys were set up to type commands like RUN, LIST, SPRDEF, GRAPHICS and others, but the user could change this using the KEY command in either direct mode or under program control.

2.52.5 Technical issues

All the versions of BASIC on Commodore's 8-bit machines used a scheme of replacing BASIC keywords with single-byte code -- e.g. the word "PRINT" would be substituted by a single byte value, or *token*, rather than the five ASCII-codes for the five letters in the word. Super Expander added more commands than this system could accommodate, so a system of two-byte tokens for the new commands was implemented.

Because of a quirk in the BASIC interpreter (the handling of the "THEN-part" of an IF/THEN construct didn't jump through a vector in RAM but instead took a direct JuMP to the standard, unexpanded BASIC command decoding routine), IF/THEN statements needed to have an extra colon (:) inserted right after the THEN keyword, if the following command was one of Super Expander's non-standard BASIC keywords. For instance, IF (condition) THEN DRAW ... would yield a ?SYNTAX ERROR message — one had to write it like IF (condition) THEN:DRAW ... — note the colon between THEN and the Super Expander-added command DRAW.

Sources

The manual is available online as eText -- http://project64.c64.org/hw/se64eng.txt

2.53   SuperCPU

The **SuperCPU** is a processor upgrade for the Commodore 64 and Commodore 128 personal computer platforms. The SuperCPU uses the W65C816S 8/16 bit microprocessor.

2.53.1   History

It was developed by Creative Micro Designs, Inc and released in 1996.[1]

It used a device called the RamCard to increase its capabilities. The card is no longer sold by Creative Micro Designs since 2001, the distribution was taken over from 2001 to 2009 by the U.S. company Click Here Software Co., but it's unclear if any were manufactured after 2001.

2.53.2   Technical description

The unit can have up to 16 MB RAM installed. The unit sported a "Turbo" switch which, when enabled, clocked a Commodore 64 or Commodore 128 up to 20 MHz. The unit plugs into the expansion port of the computer.[2] The unit uses 0.4 A and shadow ROM in 128 kB of RAM. Internal ROM was 128 kB.[3] Using the RamCard (SuperCard), Fast page (FPM) not EDO memory modules with PS/2-SIMM socket can be used in the capacities of 1, 4, 8 or 16 MB.[4]

2.53.3   External links

- The SuperCPU Home with FAQ and software.

- archive.org/geocities.com: CMD Product review, CMD SuperCPU (from 2008)

2.53.4   References

[1] "SuperCPU coding competition for 2006. - Commodore 64 (C64) Forum". lemon64.com. 2006-02-26. Retrieved 2013-05-26.

[2] "The Unofficial CMD Homepage - SuperCPU". cmdweb.de. 2010-04-18. Retrieved 2013-05-26.

[3] "SuperCPU   General   Specifications".   webcache.googleusercontent.com.   1996.   Retrieved 2013-05-26.

[4] "The SuperCPU FAQ". supercpu.cbm8bit.com. 2013-01-01. Retrieved 2013-05-26.

2.54   The Final Cartridge III

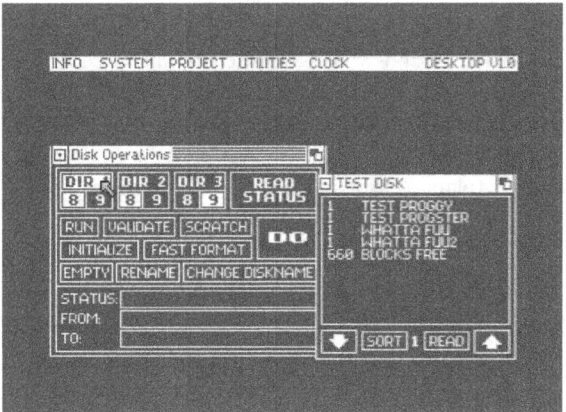

The disk utility of The Final Cartridge III GUI, showing a disk's directory

The **Final Cartridge III** was a popular extension cartridge which was created for the Commodore 64 and Commodore 128, produced by the Dutch company Riska B.V. Home & Personal Computers. It offered a fast loader, increasing the speeds of the disk drive, and a *freezer*, allowing the program execution to be stopped to be resumed later.

2.54.1   Features

The cartridge featured a "reset" button and a "freeze" button, as well as a LED that indicated whether or not the module was active. The cartridge featured a "Final Kill" option (accessible through the desktop, freezer or BASIC) which disabled the cartridge's functionality and booted the computer to unexpanded state. This was needed as some software, particularly games, were incompatible with fast-loaders; disabling the cartridge meant it never needed to be removed.

One of the unique features of the cartridge was its GUI, even when its usefulness remained quite limited compared to other GUI environments for Commodore 64. Unless RUN/STOP key was held down during power-on or reset, the cartridge presented a graphical WIMP desktop. The graphical look of the desktop was borrowed from AmigaOS 1.x. It was possible to load new GUI-based utilities from disk or tape, though these remained rare. Of the tools in the cartridge ROM, the most useful were a text editor, a disk file management utility, a calculator, and an alarm clock.

A big selling point was the disk and tape turbo feature, which was available for most commands; this accelerated loading things from disk or tape considerably. However, the biggest strength of this particular cartridge for the Commodore 64 lies in the built-in machine code monitor program, which is capable of the widest range of features, such as text and sprite dump, as well as text and sprite editing.

The cartridge provided an extension to the Commodore BASIC, which contained several new BASIC programming aids, such as RENUMBER, and several utility commands, one of the most notable of which was DOS" which can be used to give Commodore DOS commands (e.g. DOS"S0: UNDESIRED FILE to delete a file), read the error status of the drive (plain DOS") or display the disk directory without overwriting the BASIC program in the memory (DOS"$). The BASIC commands also allowed to return to the GUI desktop mode, or start the machine-language monitor.

The freezer feature allowed to save the memory contents to disk to be resumed at later point (this allowed for convenient copying of single-load games, for example). It also allowed the use of some rudimentary game cheating functionality (disabling sprite collision detection, for example), and printing a copy of the screen image to the printer. The freezer also allowed access to the machine-language monitor.

The FC3 can be combined very well with a SD2IEC-Floppy, because his Fastloader stays alive in File-Browsers like CBM or SD2BRWSE. Also you have the very good FC3 F-key commands for changing directories on the SD-Card or loading files without the need of typing anything. Therefore the combination FC3 and SD2IEC complement each other very well.

2.54.2 See also

- ISEPIC

- Trilogic Expert Cartridge

2.54.3 External links

- Project64 etext of The Final Cartridge III manual (index page, file number 33 on the list)

- Photos and Information about the Final Cartridge III

2.55 The Judges (demogroup)

The Judges was a Dutch Commodore 64 group from Roosendaal known for being one of the earliest dedicated demogroups. The Judges released several demos for the Commodore 64 home computer between the years 1986 - 1988.[1] Groups such as The Judges are regarded as early pioneers of what came to be known as the demoscene.[2]

In some respects they mirrored The Lords, a "rivaling" ZX Spectrum group also from Roosendaal, to the extent that concepts, artwork and even titles were identical between the two groups.

A technical milestone often attributed to The Judges, particularly the programmer Bart "White" Meeuwissen, is the invention of the FLD (Flexible Line Distance) technique,[3] which was used in the *Think Twice* series of demos. As the name implies, FLD made it possible to have a variable distance every eight pixels between the individual text or graphics lines generated by the VIC-II video chip.

The group's musician Jeroen "Red" Kimmel went on to compose and sell video game music commercially for different platforms such as the C64, Amiga and MSX computers.[4]

2.55.1 Members

- Jeroen "Red" Kimmel

- Bart "White" Meeuwissen

- Hans "Der Hansie" van Gink

- Corne "Coko" Koen

2.55.2 Releases

- Think Twice series (parts 1-5, released in 1987 and 1988)

- Crazy Sample 1 & 2

- Hubbard Track series (parts 1-3)

- Rhaa Lovely 1 & 2

- Touch Me

- Jugglin' Judge
- Rascal
- Mikie's Music
- It's a Kind of Magic
- Phantom of the Asteroid Music

2.55.3 References

[1] The Judges at pouet.net

[2] Copyright Does Not Exist, Chapter 5 - Subculture of the Subcultures

[3] FLD - Plus 4 Encyclopedia - Plus 4 World

[4] Jeroen Kimmel Video Game Credits and Biography - MobyGames

2.55.4 External links

- Information page at CSDB
- Some demos
- Interview of White at C64.COM
- The Judges at pouet.net

2.56 Trilogic Expert Cartridge

The **Expert Cartridge** from the UK company Trilogic was a popular extension cartridge which was introduced in mid-1986 for the Commodore 64 and a later Commodore 128 compatible version, It offered a *fast loader*, increasing the speeds of the disk drive, and a *freezer*, allowing the program execution to be stopped to be resumed later. Later Cat & Korsh International took over the distribution and the Trilogic name disappeared from the cartridge.

A major difference to the other two The Final Cartridge III and Action Replay is that it held its system software in an 8 kB RAM that could be reprogrammed. And thus allowed the user to change its functionality.

The author of the Expert cartridge default firmware is John Twiddy who also wrote the games The Last Ninja and Ikari Warriors for the Commodore 64. He claims that it saved him many months of programming.[1]

2.56.1 See also

- ISEPIC - An earlier cartridge with 2 kB RAM
- The Final Cartridge III

2.56.2 References

[1] Your Commodore Issue 48, September 1988

2.56.3 External links

- c64.org - Expert Cartridge

2.57 Ventilator 202

Ventilator 202 (in Serbian language, meaning: "Electric fan" 202) was a live radio show broadcast by Beograd 202 radio station during the 1980s and hosted by Zoran Modli. It was one of the most important shows of Belgrade's "202" station and possibly also the most important project of its host. It first aired first June 3, 1979. Zoran Modli was its host until late 1987. He later hosted another similar show, Modulacije. "Ventilator 202" was renamed to "501" and hosted by Dubravka Marković, giving it her own style. It was notable for its promotion of local (domestic) demo music, early application of computers and introduction of "absolute radio" concept.

Ventilator 202 was a show done differently from other contemporary radio shows in Yugoslavia. Zoran Modli was not only the show's host but also operated the mixing console and other equipment himself. Essentially he was also a disc jockey.

During the second half of the 1980s, Ventilator 202 broadcast computer software recorded on cassette tapes for popular home computers Galaksija, ZX Spectrum and Commodore 64.

2.57.1 Demo Music

A number of other contemporary radio shows broadcast demo music made by young bands but failed to gain popularity. Zoran Modli took a different approach with Ventilator 202. Instead of playing solely demo music he interleaved it with hits of the time, as equal. The show gained such popularity that it became a great starting point for young musicians and a way to hear an alternative to established sound. Ventilator 202 had no shortage of demo tapes coming in, from the entire Yugoslavia of the time.

The popularity eventually lead to the idea that actual albums (vinyl records and cassette tapes) can be released with the best of the best. With the help of Dušan Pančić, Ventilator 202 released the first "Demo Top 10" album in April 1983, soon to be followed by a second one in November 1983. The last, third, album created by the original team was released Spring 1985.

Volume 1 - April 1983

- Rex Ilusivii: "Zla kob"

- Šta se vidi kroz durbin: "Želi da ga zaboravi"

- Partibrejkers: "Radio utopija"

- Berliner Strasse: "Maske"

- Miško Plavi: "Hemija"

- Petar i Zli vuci: "Moroni"

- 39. legija: "Smisao"

- Zak: "Kosmička balada"

- Laki Pingvini: "Možda, možda"

- Via Talas: "Ti"

Volume 2 - November 1983

- Ajfel na kraju: "Mahatma"

- Atlantski poremećaj: "Ništa se ne dogadja"

- Tadaima: "Poželih"

- Belo belo: "Ružičasta bluza"

- Aja Sofija: "Poslednji mangup sa Dorćola"

- Aja Sofija: "Uvek si tu"

- Amerika snova: "Andaluzija"

- Vatrostalno staklo: "Ovih dana nije smela"

- Vanila pakt: "Andaluzija"

- Gustaph i njegovi dobri duhovi: "Upotreba majmuna"

Volume 3 - Spring 1985

- Rex Ilusivii: "Arabia"

- Fleke: "Slatka mala"

- Karlowy Vary: "Ratnici"

- Sayonara: "Neka draga lica"

- La Strada: "Želje"

- Ursula i provincijalac: "Budilnik za Radmilu P."

- Belo belo: "Tamni vilajet"

- Psihomodo Pop: "Zauvjek"

- Oskarova fobija: "Beli dekolte"

- Solunski front: "Ratnom drugu čast"

- Mrgudi: "Proleteri"

2.57.2 Broadcasting Computer Software

The show went as far as broadcasting software for computers of the day, such as Galaksija, ZX Spectrum and Commodore 64. The host would announce the details of the software about to be broadcast (title, type of computer, etc.). Ready listeners would then engage their tape recorders and record the characteristic noise that would have turned unsuspecting listeners away.

This initiative became so popular that special software was made specifically for the show, such as the flight simulator and "Velika Akcija" (Great Action) games and an early electronic journal. The journal was broadcast as the entire program used to view it.

Ventilator 202 also featured computer professionals and publicists talking about various interesting subjects. Frequent speakers were Vladimir Ajdačić, Miša Milosavljević, Gavrilo Vučković, Jovan Regasek, Tansije Gavranović, Zoran Živković, Aca Milinković and Dejan Ristanović.

2.57.3 Ventilator 202 team

- Zoran Modli, host and DJ

- Dušan Pančić

- Vladimir Ajdačić

- Miša Milosavljević

- Voja Antonić

- Dejan Ristanović

2.57.4 See also

- Ventilator 202 - article written by Zoran Modli, the show's host. In Serbian language.

- Nicest demo recordings from "Ventilator 202" - a collection of MP3 and RealAudio samples of demo songs broadcast by Ventilator 202, maintained by Zoran Modli, the show's host. In Serbian language.

- Games published by Ventilator 202 - a list of games developed and broadcast by Ventilator 202 team.

- Ventilator 202, *Blic*, October 16, 2008

2.58 Xetec

Xetec /ˈziːtɛk/ was founded in 1983 by Jon Flickinger, and was located in Salina, Kansas, United States. Before closing in 1995, the company produced many third-party products for the Commodore 64, Commodore 128, Amiga, Macintosh, Atari ST and PC computers.

2.58.1 Overview

The Lt. Kernal is the first third-party hard drive peripheral for Commodore computers. Originally developed by Fiscal Information in 1985, it was turned over to Xetec for manufacturing and customer support.[1]

Xetec's best selling product is its line of printer interfaces, which allows the use of many models of non-Commodore parallel-interface printers with Commodore computers. Some of the popular printers of that era that are supported include Canon, C-Itoh, Star Micronix, Epson, NEC, Okidata, and Panasonic. Early interface models (such as the "Serial Printer Interface" and "Graphics Printer Interface") simply convert from Commodore's serial format to the more standard Centronics parallel interface, with only minimal ASCII conversions and graphic character printing. The Super Graphix Jr adds support for 50+ printers and "Near Letter Quality", which is a technique of using multi-pass graphic printing to achieve higher quality text printing. The more sophisticated Super Graphix also adds an 8K data buffer, screen dump support, two user-loadable fonts (from a library of fonts on the included disk), and a font creation program. The Super Graphix Gold adds a 32K buffer, 4 fixed and 4 user-loadable fonts (from a library of fonts on disk), 10 font printing effects, picture printing, built-in screen dump programs, fast-serial support, and the rather unique ability to interface a printer directly to a disk drive (for printing pictures and loading fonts directly from disk).[2]

The Fontmaster line of word processing software provides Commodore users the ability to exploit the graphics capabilities just emerging in printers of the day in order to produce documents containing a mixture of font styles, sizes and effects as well as embedded pictures. Although lacking the power and flexibility of word processors available today, it was ground-breaking in its day, winning an *Outstanding Original Programming* award at the 1985 Consumer Electronics Show. Fontmaster 128 was featured in the 1987 Consumer Electronics Show Software Showcase.

Xetec's offerings for the Amiga include SCSI interfaces, hard drives, CD-ROM drives, streaming tape drives, and RAM expansion. Xetec was the first to offer the popular Fish disks collection on CD-ROM with their three-volume set of *Fish & More* discs.

Xetec also developed a few products for Macintosh and PC computers, products for the RC hobby market, and spent a number of years in research and development of solid-state

fluorescent lighting ballasts, for which Jon Flickinger holds two patents.[3][4]

In summary, Xetec's total list of products include these: Lt. Kernal, Serial Printer Interface, Graphics Printer Interface, Super Graphix, Super Graphix Jr, Super Graphix Gold, Graphix AT, Fontmaster, Fontmaster II, Fontmaster 128, Printer Enhancer, FastTape, FastTrak, FastCard, FastCard Plus, MiniFastCard, FastRAM, *Fish & More* (vols. I, II, III), CDx CD-ROM Filesystem, CDx 650, Beeping Tomm, SuperWriter 924, SCAD.

2.58.2 References

[1] Lt. Kernal history

[2] Scanned Xetec user manuals

[3] Flickinger, Jon. "Patents by Inventor Jon Flickinger". *Justia*. Retrieved August 22, 2014.

[4] Flickinger, Jon. "Series resonant inverter and method of lamp starting". *Google*. Retrieved August 22, 2014.

2.58.3 External links

- Lt. Kernal support site

- Amiga Hardware Database - Xetec Amiga expansion cards

- The Big Book of Amiga Hardware - Photos of Fast-Card

- The Big Book of Amiga Hardware - Photos of Fast-Trak

- Review of CDx CD-ROM FileSystem

- ANTIC VOL. 6, NO. 8 / DECEMBER 1987 / PAGE 54 - ANTIC's mention of SCAD

- Easter eggs in Super Graphix Gold

- Easter eggs in Printer Enhancer

- Graphix AT (Atari) review

Chapter 3

Text and image sources, contributors, and licenses

3.1 Text

- **Commodore 64** *Source:* https://en.wikipedia.org/wiki/Commodore_64?oldid=686649129 *Contributors:* Damian Yerrick, Tuxisuau, Robert Merkel, The Anome, Berek, Tarquin, Wayne Hardman, Ben-Zin~enwiki, Cyrek, Imran, Stevertigo, Frecklefoot, Edward, Ubiquity, Ceaser, Smelialichu, Michael Hardy, Alodyne, Norm, Mahjongg, Tenbaset, Nixdorf, Liftarn, Wwwwolf, Dave Farquhar, Penmachine, Tregoweth, Looxix~enwiki, Samuelsen, Julesd, Poor Yorick, BlueEel, Bavi H, Grin, Harvester, Smack, GRAHAMUK, Conti, Malcohol, Slark, Whisper-ToMe, Rvalles, Gestumblindi, Maximus Rex, Furrykef, Grendelkhan, Wernher, Thue, Topbanana, Raul654, Jeffq, Rossumcapek, Twang, Zeke (usurped), Fredrik, Korath, Boffy b, RedWolf, Dittaeva, Psychonaut, Blainster, Gwalla, JamesMLane, DocWatson42, Jonth, Massysett, Lupin, Ds13, Everyking, Bkonrad, Guanaco, Sundar, AlistairMcMillan, Foobar, Bobblewik, Golbez, Junkyardprince, Lvr, SebastianBreier~enwiki, Toytoy, Quadell, Beland, Kaldari, Jossi, Murple, Nzseries1, Bumm13, Jawed, Necrothesp, Eranb, Lumidek, B.d.mills, Urhixidur, Marcus2, Joyous!, Oknazevad, Tibi, Grm wnr, Fuzlyssa, Chmod007, Avatar, TheCustomOfLife, Ta bu shi da yu, Jiy, Jpg, Discospinster, Rich Farmbrough, Guanabot, Sladen, Loganberry, Pak21, Qutezuce, Jpk, Pixel8, Ivan Bajlo, Ponder, Blakespot, Indrian, Horsten, Bender235, Speedbump, Djordjes, Dcabrilo, Ylee, CanisRufus, Pjf, Richard W.M. Jones, Chairboy, RoyBoy, Femto, Incognito, Dgpop, Fugazi32, DanielNuyu, Deathawk, Mike Schwartz, Duk, NotAbel, Simonroy, Matt Britt, Vintermann, Mr2001, Diceman, SpeedyGonsales, Tyan23, Fotinakis, Spiro Trikaliotis~enwiki, Estrus, Datucker~enwiki, Frodet, Alansohn, Borisborf, Tablizer, Duffman~enwiki, Miranche, Jamyskis, Guy Harris, Fadookie, Halsteadk, Interiot, Borisblue, Neonumbers, CyberSkull, M7, Dowcet, Ashley Pomeroy, InShaneee, Bart133, GeorgeStepanek, Gbeeker, Wtshymanski, Almafeta, Suruena, Garzo, Jheald, Drat, Parody, Gortu, Feline1, Voxadam, New Age Retro Hippie, Blaxthos, Dismas, Dtobias, Mmarkley, Mcsee, Preaky, Anttivs, Woohookitty, LOL, Ae-a, Guy M, Commander Keane, Cscott, Kelisi, Hailey C. Shannon, Bennetto, Al E., Dah31, Radhe2k, Mb1000, ThomasHarte, Daniel Lawrence, Xiong Chiamiov, Marudubshinki, Behun, Virtualsky, Graham87, Magister Mathematicae, Qwertyus, SamuraiClinton, FreplySpang, JIP, Ketiltrout, CelticWonder, JoshuacUK, MZMcBride, ErikHaugen, SMC, Vegaswikian, Funnyhat, Oblivious, Mecandes, SeanMack, Bubba73, GregAsche, Husky, SLi, The Deviant, StuartBrady, FlaBot, Ianthegecko, Mirror Vax, SchuminWeb, RobertG, KarlFrei, SuperDude115, Jelans2001, Ewlyahoocom, Swtpc6800, Davetron5000, Crazeman, Arickp, Zotel, Jared Preston, Random user 39849958, VolatileChemical, ShadowHntr, Niz, Metaeducation, The Rambling Man, Wavelength, JJLeahy, Crotalus horridus, Todd Vierling, OtherPerson, RussBot, Petiatil, Dleonard, Fabartus, Silverdr, Gateman1997, Wgungfu, Bovineone, Wimt, Thane, NawlinWiki, Dforest, King V, 9cds, Cleared as filed, Retired username, Jeffme, CPColin, PhilipO, Skl1983, Zwobot, Mysid, Tescomarc, Cjs, Dartiss, Judacris, Richardcavell, Rbirkby, Glome83, Penguinpc, Musicomputer, NotInventedHere, Natgoo, JLaTondre, Fraxen, Katieh5584, Dzfoo, Kazmeyer, Jeff Silvers, Mewcenary, Alebed01, Kiousu, Mvdwege, AndersL, SmackBot, MattieTK, Hux, Nadix 1, Pgk, Jagged 85, Nidnid, Isenguard, AnOddName, Geoff B, HalfShadow, Oscarthecat, Skizzik, Carambola, Bluebot, Flurry, Kurykh, BullWikiWinkle, Tghe-retford, Sirex98, Thumperward, Chrown~enwiki, Conway71, OrangeDog, Apple2gs, Ctbolt, MichaelWheeley, ToobMug, Konstable, Nintendude, Emurphy42, Theneokid, Rrelf, J00tel, Dethme0w, OrphanBot, Mr.bonus, Nixeagle, Cybercobra, Paul Panks, EVula, Michaeljgrasso, DaveMMR, Voss749, Sigma 7, Tsalsa, Deepred6502, Nishkid64, John, Guroadrunner, ML5, Tpulak, Gang65, IronGargoyle, Farazparsa, Omnedon, Nissefoo, James@videogamejunkie.co.uk, TastyPoutine, Hetar, IvanDíaz, Drvanthorp, Blehfu, Courcelles, Marcus Bowen, Clyde Miller, Vaughnstull, WolfgangFaber, Nutster, SkyWalker, J Milburn, CmdrObot, Jlbarron, Washi, Jedgeco, Ninetyone, Scirocco6, DevinCook, Nczempin, GerryJJ, Cydebot, Farine, MC10, Prodigyman1970, Bluedevil04, Gogo Dodo, Dancter, Tawkerbot4, DumbBOT, Jay32183, NorthernThunder, StoneGiant, Arb, Gimmetrow, Stanleylieber, Thijs!bot, JAF1970, Zornfalke, Kubanczyk, Drarkanex, Stonic, DavidLaas, Marek69, Electron9, WinstonSmith6079, Silver Edge, Big gun, Bil Herd, AntiVandalBot, Luna Santin, Guy Macon, Seaphoto, KickahaOta, Moorematthews, Bladestorm, Jhsounds, DOSGuy, JAnDbot, DuncanHill, ThomasO1989, MER-C, Sjlain, Revener, Longstreet87, Acroterion, Geniac, Wasell, Nicolaasuni, Wikilolo, Meeples, Drugonot, BMW Z3, Retrometro, Ecksemmess, Robotman1974, Nreive, Unsigned Char72, Adriaan, Vanessaezekowitz, Zeus, Ulkomaalainen, CommonsDelinker, Chronodev, Nono64, Radbug, Doctorcasey, Tom Paine, Lozzaaa, Richiekim, Fierman, Uncle Dick, Mdmeyer, Alexanda, Borisser, Sega31098, Adamjansch, Dispenser, Jeepday, Grez868, Wilderns, Plasticup, Dcmacnut, Bigdumbdinosaur, DMCer, Scott Illini, Commander-64, CardinalDan, Tomhollander, VolkovBot, Morenooso, Leopold B. Stotch, TCZ, Katydidit, Mike Yaloski, WOSlinker, TXiKiBoT, KonstableSock, GimmeBot, Sarenne, Gerrish, Jimbowp, Corvus cornix, GeneralBelly, Akaimaru, Speccyclive, Jonnyjack, Gona.eu, WinTakeAll, Ianjones50, Andy Dingley, Falcon8765, Bunnyhugger, Austriacus, Wjl2, Fnagaton, Randommelon, Spikey Meister, Nestea Zen, RJaguar3, Jerryobject, Mr.Z-bot, Knowocean~enwiki, Theaveng, DECM, Snideology, TheMasterDude, Josso (usurped), C'est moi,

JohnnyMrNinja, ThisGuy62, Echo95, Superbeecat, Illinois2011, ImageRemovalBot, Faithlessthewonderboy, Martarius, Sfan00 IMG, ClueBot, Hippo99, Badger Drink, The Thing That Should Not Be, Infectiousconcepts, MikeVitale, Iuhkjhk87y678, Niceguyedc, Flaming, Petestar1969, Excirial, Eeekster, Muhandes, Sun Creator, Dekisugi, A plague of rainbows, Le grande banane, PCHS-NJROTC, Ubardak, DumZiBoT, Final-night, Tobias.hultman, XLinkBot, Prof Wrong, CheShA, Petchboo, Ost316, Monkthatgotfunk, Colliric, Gamernotnerd, Curtlee2002, Addbot, Mortense, Magus732, M.nelson, Stentie, Ronhjones, RedRose333, Calle1970se, Download, LemmeyBOT, Fireaxe888, Tide rolls, Teles, Et-trig, Swooboo, Softy, Werewolf Bar Mitzvah, Legobot, Yobot, II MusLiM HyBRiD II, Pcap, StationHub, AnomieBOT, Pandorym, RBM 72, CraigBox, Law, Materialscientist, Philsan, Snorlax Monster, Boxstaa, Ayda D, Hexadecima, Xqbot, APDzie, Fgroover, Capricorn42, Jeffrey Mall, Wastedinamerica, John3790, Ubcule, Sir Stanley, GrouchoBot, Kyng, Chaheel Riens, Sesu Prime, FrescoBot, Surv1v4l1st, Haeinous, Lyscarz, Rectec794613, LittleWink, Smuckola, Jaguar, Salvidrim!, Full-date unlinking bot, Cnwilliams, Ajith Kumar KV, Plbyrd, SchreyP, Commandr Cody, Matthewalborn, Aisha9152, Fatgreeny, PleaseStand, Dtgriffith, DARTH SIDIOUS 2, Betateschter, Wirepath, Gahpublic, John of Reading, Kool kitty89, Zollerriia, Dewritech, Lapsklaus, Rthiebaud, Slightsmile, Zictor23, Cogiati, Hydao, Dolovis, Finderetrex, Finemann, Bobjohnsonwalmart, Jonpatterns, Joshtipka, Wayne Slam, Taylorisfat, Sbmeirow, Δ, Retrogaming, Wikiwriter80132, Donner60, RRabbit42, Avivanov76, Bill Hicks Jr., ChuispastonBot, Evan-Amos, Targaryen, Petrb, Faramir1138, ClueBot NG, Pantergraph, Bobjeats, Kylemstinks, Chester Markel, IlRoberto88, 8bitdemo, Helpful Pixie Bot, Compilation finished successfully, Wbm1058, Lowercase sigmabot, Sunzuki123, BG19bot, Mb777, Strobnostification, Tilmen, NukeofEarl, Gravitoweak, What would YOU like to be?, RadicalRedRaccoon, JDNB, BattyBot, NewExLionTamer, Cyberbot II, ChrisGualtieri, NeatlyTiled, StevefromQuebec, Dexbot, Mjb1962, Makecat-bot, ABunnell, MrHealesville-VIC3777, BDE1982, SFK2, C64hater, Altontowers123, Condorcraft110, Miller9904, Eyesnore, Commodore 63, Kolekostur, Retroisler, Dude-WithAFeud, JaconaFrere, PaulLondoner, Monkbot, Smashbot64, Haasid, TheKingsTable, OMPIRE, Chambr, SummerFunMan, Falloutah, Gamingforfun365, Jjb1017, KasparBot, Were You There Too?, So-retro-it-hurts, CyberWarfare and Anonymous: 736

- **1541 Ultimate** *Source:* https://en.wikipedia.org/wiki/1541_Ultimate?oldid=585271687 *Contributors:* Electron9, Addbot, Khazar2, Alphamyon and Anonymous: 2

- **Action Replay** *Source:* https://en.wikipedia.org/wiki/Action_Replay?oldid=674865047 *Contributors:* Liftarn, Wwwwolf, Ahoerstemeier, Net-snipe, Silverfish, Maximus Rex, Furrykef, Wernher, PxT, Matt Gies, NessSnorlax, Rick Block, Gilgamesh~enwiki, Dave2, Zikar, JoJan, Abdull, Zoganes, Mattb90, Discospinster, Guanabot, Pixel8, AndrewM1, Evice, Ht1848, CanisRufus, Bobo192, Bttfpromo, Kappa, Pschemp, Jason One, Alansohn, Gary, Jtalledo, Ricky81682, Seancdaug, ReyBrujo, ThatNateGuy, Marasmusine, Cruccone, Al E., ThomasHarte, Transfa-tex, Xizer, RadioActive~enwiki, Seventh Holy Scripture, Jorunn, Rjwilmsi, Commander, Ikh, Yamamoto Ichiro, Ian Pitchford, Mirror Vax, PinkDeoxys, King of Hearts, Bgwhite, Agamemnon2, YurikBot, LittleSmall~enwiki, Retodon8, RussBot, Muchi, DarkfireTaimatsu, Mark-cox, Gaius Cornelius, TonicBH, Matticus78, Jeffpower, VederJuda, Wknight94, NorsemanII, Viande hachée, Maxamegalon2000, Rwwww, SmackBot, State of Love and Trust, Jagged 85, Gilliam, Ohnoitsjamie, Oscarthecat, Gaiacarra, 32X, Missingno000, Iamstillhiro1112, Raymie, Colonies Chris, Can't sleep, clown will eat me, Blah2, OrphanBot, Irish Souffle, Zmon1, Richard0612, 汪汪汪, Juux, Rockvee, Pyksy, MarkSutton, Anton10000, The Tsar, Optakeover, BrownCow, RememberMe?, JLukas, Moped, Ryulong, Phuzion, TJ Spyke, JoeBot, J Di, Az1568, Radi-ant chains, ChrisCork, Lhasapso, Information Center, Jesse Viviano, Rmallins, Dogman15, Cydebot, Thaddius, Anpu777, SJ2571, Blindman shady, BetacommandBot, Epbr123, Cosmi, Kenshiroh, Tapir Terrific, X201, Cool Blue, Dfrg.msc, AntiVandalBot, Dude902, OptimumTaurus, Jhsounds, 55david, Green Hill, Armando12, Tony Myers, MER-C, Baby Luigi, PhilKnight, Lemmayoshi2, VoABot II, Marko75, Kakomu, CameronB, Legasafe, Imagineallthepeople92, Cander0000, DerHexer, Ugetab, HehEXE, Pikazilla, MartinBot, PAK Man, OmegaStriker, King Sweaterhead, Tgeairn, J.delanoy, Uncle Dick, TheomanZero, Bloodrain~enwiki, Halo123spartan, Derlay, Katalaveno, LordAnubisBOT, Samthe-boy, AntiSpamBot, Budster650, Xeysz, Ace of Jokers, Cometstyles, Casper10, Mash1972, Durdy, Glossologist, Lights, 28bytes, Fbifriday, Floppydog66, Gamer313, Dachi, Rei-bot, Qxz, MiracleMatter, Martin451, Maxim, Complex (de), Haseo9999, Victory93, Brianga, Nagy, Tv-inh, SieBot, Winchelsea, One more night, Max A K Challie, Sp4rt4n, Proud Ho, JetLover, Oxymoron83, Faradayplank, Techman224, Zixor, JohnnyMrNinja, Altzinn, L.C.E.C., Martarius, ClueBot, LAX, EoGuy, Mild Bill Hiccup, Niceguyedc, WikiMesser, Resoru, Abrech, Spark-dogvbdapdnw, Elizium23, Pianokid54, Aj00200, Feinoha, Sully76cl, Lstanley1979, WikHead, Newbiez, Subversive.sound, SelfQ, Addbot, Benito2, Willking1979, Tcncv, GeneralAtrocity, Joshual2153, MrOllie, The First Darklord, Tassedethe, Gail, HerculeBot, Luckas-bot, Yobot, Terrifictriffid, Skyress1, Tempodivalse, Wiki Roxor, AnomieBOT, 1exec1, Piano non troppo, Flewis, LilHelpa, Xqbot, Belasted, Addihockey10, Junkcops, Tad Lincoln, Abdul qayyum 1986, Brandon5485, Chaheel Riens, Link83, Jasonjambalya, Drew R. Smith, Dude1818, TimmyX, Wi-iKiBoyz, Edituser7110, Gonefishin12, John of Reading, Tuankiet65, Mysteryman19, GoingBatty, Jjesseshort, RibShark, AlphaPikachu578, Ilelac, Jsayre64, Megagboy1, Guillaume0320, Fredy360, ClueBot NG, JasonRandom2012, Marechal Ney, ARCodes, Ryanteoh, Gorgak25, DMStern, Ghosty567, Power Macintosh, Fylbecatulous, RichardMills65, Rusesji, Carts9999, Origamite, Earthsea98, Prisencolin, Rdpfluke and Anonymous: 412

- **C-One** *Source:* https://en.wikipedia.org/wiki/C-One?oldid=628967722 *Contributors:* Mahjongg, Pnm, Pwsoft, Wernher, Toytoy, Houshuang, Damieng, Sysy, Polluks, Magetoo, AndrewWatt, GregorB, MagerValp, Mirror Vax, Stoph, YurikBot, Nick, SmackBot, Elonka, Thumperward, Appaloosa2k, Frap, Alast0r, ChrisCork, JLD, Al Lemos, Marko75, Wikip rhyre, JRS, Andy Dingley, Addbot, Lightbot, Yobot, Lemmy Laffer, Xorxos, Shaddim, Khazar2 and Anonymous: 18

- **C64 Direct-to-TV** *Source:* https://en.wikipedia.org/wiki/C64_Direct-to-TV?oldid=680267677 *Contributors:* Gbraad, Mahjongg, Tregoweth, Schneelocke, Wernher, Morn, Psychonaut, Brian Kendig, Everyking, Chowbok, Toytoy, Bumm13, Tubedogg, Sysy, TedPavlic, R.123, Si-etse Snel, Polluks, Gargaj, Yair reshef, Magetoo, Sobolewski, ReyBrujo, C3o, MagerValp, HiFiGuy, JIP, Rjwilmsi, Bensin, Mirror Vax, D.brodale, YurikBot, Dialectric, Amigan, NotInventedHere, Fourohfour, Eenu, SmackBot, Thumperward, Can't sleep, clown will eat me, SanderK, Huax~enwiki, Fekker, Fred nd, Mblumber, Peteb16, Lugnuts, Pyrogen, Thijs!bot, Electron9, Aruffo, The Wizard Guy, JAnDbot, STBot, DeeKay64, Cmdr Scolan, Gruverja, Thunderbird2, Fnagaton, Ferret, Juandope, Felix the Hurricane, DumZiBoT, InternetMeme, Ad-dbot, JBsupreme, Incog88, Ganimoth, AnomieBOT, Danno uk, Lawrence Davis, Glider87, CrazyIcecap, Updatehelper, GoingBatty, ZéroBot, Clockwork Hydra, Targaryen, Cartakes and Anonymous: 44

- **CARDCO** *Source:* https://en.wikipedia.org/wiki/CARDCO?oldid=645810089 *Contributors:* Rich Farmbrough, Fastily, Dsimic, Sbmeirow, Mhiji and FSII

- **Cc65** *Source:* https://en.wikipedia.org/wiki/Cc65?oldid=668932964 *Contributors:* Damian Yerrick, Tannin, Wwwwolf, Furrykef, Itai, Wernher, Psychonaut, Stewartadcock, Dbenbenn, Bumm13, Polluks, Mirror Vax, RussBot, Ntsimp, DmitTrix, Gwern, FrescoBot, EdoDodo, KLBot2, ChrisGualtieri and Anonymous: 14

- **CMD RAMLink** *Source:* https://en.wikipedia.org/wiki/CMD_RAMLink?oldid=638340567 *Contributors:* Wernher, Klemen Kocjancic, Mirror Vax, CWenger, SmackBot, Vanessadannenberg, Gang65, Thijs!bot, Electron9, Erechtheus, Vanessaezekowitz, Imperator3733, Niceguyedc, Suenarmy, Addbot, Andi boe, FrescoBot, John of Reading and Anonymous: 4

- **Commodore 1520** *Source:* https://en.wikipedia.org/wiki/Commodore_64_peripherals?oldid=685484408 *Contributors:* Wernher, Tomlouie, Boffy b, Altenmann, Marcika, Sdfisher, Uzume, Bumm13, Pixel8, Polluks, Nholland~enwiki, Kelly Martin, Firsfron, Woohookitty, Bellhalla, Dah31, Shadyman, BD2412, JIP, Rjwilmsi, Vegaswikian, Dbsanfte, Ian Pitchford, Bgwhite, Digitalme, Crotalus horridus, RussBot, Silverdr, Nikkimaria, Scolaire, SmackBot, Betacommand, Oscarthecat, Chris the speller, Bluebot, Bazonka, Vanessadannenberg, CSWarren, Colonies Chris, Emurphy42, Bogsat, BlackTerror, John, Gang65, Dodgem4s, TastyPoutine, DagErlingSmørgrav, CmdrObot, DevinCook, Astralblue, Neelix, Mblumber, Jaktrip, Marksmithnz, Uspn, Trevyn, Electron9, Monkey Man, Messeng3r, T@nn, Destynova, Stephenchou0722, CommonsDelinker, Nono64, PyroGuy, LordSkitch~enwiki, Multicherry, DeeKay64, Bigdumbdinosaur, Soci64, Sarenne, Somnix, Merman1974, Haseo9999, Fnagaton, Rlendog, Lightmouse, IdreamofJeanie, MikeVitale, Mild Bill Hiccup, Crywalt, Petchboo, Ovalnovel~enwiki, Addbot, Michael480, Calle1970se, Yobot, Fraggle81, AnomieBOT, Locos epraix, Ubcule, Locobot, Babylon4, Chaheel Riens, Kwiki, Jopinder, Plbyrd, Soenke Rahn, Dexter Nextnumber, Stasvasy, Angrytoast, AvicAWB, Sbmeirow, Bomazi, Evan-Amos, Matthiaspaul, Compilation finished successfully, BG19bot, Alphamyon, Mogism, Monkbot, OMPIRE, Mbp303, Wikidalien and Anonymous: 63

- **Commodore 1541** *Source:* https://en.wikipedia.org/wiki/Commodore_1541?oldid=682489245 *Contributors:* Damian Yerrick, NathanBeach, Tarquin, Stephen Gilbert, VincentV, Wwwwolf, Dave Farquhar, Ahoerstemeier, ZoeB, Docu, Emperorbma, Crissov, Wik, Furrykef, Tempshill, Wernher, Optim, Psychonaut, Orpheus, Ds13, Guanaco, Bobblewik, Coldacid, Sam Hocevar, Rich Farmbrough, Pixel8, Ylee, JoshuaRodman, Tadman, Gargaj, Firsfron, Dah31, Knuckles, Pfalstad, Virtualsky, JIP, MZMcBride, Mirror Vax, Avalyn, Russavia, YurikBot, Crotalus horridus, Todd Vierling, Silverdr, Thiseye, Ospalh, Rallette, Elkman, Jason404, KJBracey, Frymaster, Bluebot, Kurykh, Thumperward, Can't sleep, clown will eat me, Wonderstruck, MaxxFordham, Techsmith, Washi, Luminifer, Al Lemos, Electron9, Magioladitis, Mwbrooks, Tom Paine, Reedy Bot, Themel, EoGuy, Ente75, Download, Legobot, Yobot, Afrank99, AnomieBOT, Cavarrone, LilHelpa, Ubcule, GrouchoBot, Shadowjams, Dexter Nextnumber, ClueBot NG, DieSwartzPunkt, Wbm1058, BattyBot, Bree's Block, Hydradix, Sofia Koutsouveli, OMPIRE and Anonymous: 58

- **Commodore 1570** *Source:* https://en.wikipedia.org/wiki/Commodore_1570?oldid=664565643 *Contributors:* Tarquin, Dave Farquhar, Wernher, SD6-Agent, Coldacid, Guanabot, CanisRufus, SickTwist, Mirror Vax, Sherool, YurikBot, Crotalus horridus, Museo8bits, Mgreenbe, Bluebot, Bartmc, Fayenatic london, CommonsDelinker, Db the dba, Legobot, AnomieBOT, Erik9bot and Anonymous: 5

- **Commodore 1571** *Source:* https://en.wikipedia.org/wiki/Commodore_1571?oldid=682302155 *Contributors:* Imran, Tannin, Dave Farquhar, Wernher, Coldacid, Klemen Kocjancic, Polluks, Mirror Vax, Aalegado, YurikBot, Crotalus horridus, Musicomputer, SmackBot, Slashme, Yanksox, Iridescent, Neelix, Edwing~enwiki, Thijs!bot, Electron9, Thunderbird2, Android Mouse Bot 3, Tbsdy lives, Lightbot, Legobot, AnomieBOT, DavideAlberani, WikiPhu, Bomazi, ChrisGualtieri, Hmainsbot1, OMPIRE and Anonymous: 16

- **Commodore 1581** *Source:* https://en.wikipedia.org/wiki/Commodore_1581?oldid=687123268 *Contributors:* Tarquin, Stephen Gilbert, Tannin, Dave Farquhar, ZoeB, Emperorbma, Wernher, Fedi, Coldacid, Klemen Kocjancic, Pixel8, GregorB, Graham87, Mirror Vax, Crotalus horridus, Museo8bits, Bluebot, Vanessadannenberg, Washi, Electron9, Jonay81687, RingtailedFox, Don4of4, Suenarmy, Download, Legobot, Dexter Nextnumber, Mo ainm, Faizan, Mjuxs and Anonymous: 20

- **Commodore 64 disk / tape emulation** *Source:* https://en.wikipedia.org/wiki/Commodore_64_disk_/_tape_emulation?oldid=636220970 *Contributors:* Thumperward, Avicennasis, Ssynnes, Addbot, Yobot, FrescoBot, Stasvasy, John of Reading, GoingBatty, Checkingfax, Vektorabschalter, Astrocog, Matthiaspaul, Delusion23, Erin100280, Comatmebro, Mbp303 and Anonymous: 3

- **Commodore 64 Games System** *Source:* https://en.wikipedia.org/wiki/Commodore_64_Games_System?oldid=671245875 *Contributors:* Wayne Hardman, Liftarn, Wwwwolf, Wernher, Psychonaut, Carnildo, Mitaphane, Biot, Rich Farmbrough, Pixel8, ESkog, Longhair, Diceman, Jamyskis, CyberSkull, Ianblair23, Djsasso, Woohookitty, Dah31, ThomasHarte, JIP, Isaac Rabinovitch, FlaBot, Mirror Vax, Mathrick, Stormwatch, D.brodale, Wgungfu, Th1rt3en, 2fort5r, AndersL, SmackBot, Hmains, Betacommand, J.L.Main, Thumperward, Nintendude, BlackTerror, Calysma, TJ Spyke, FairuseBot, Ronaldvd, Cydebot, Guyinblack25, Col. Hauler, Thijs!bot, Electron9, Silver Edge, Ex-Nintendo Employee, Magioladitis, R'n'B, Fierman, Rhinestone K, Guru Larry, Sarenne, Koopa turtle, Merman1974, Suriel1981, SieBot, Trivialist, DragonBot, Project FMF, WikHead, MystBot, Addbot, Luckas-bot, Yobot, Xqbot, Ubcule, FrescoBot, Sidna, EmausBot, WikitanvirBot, ZéroBot, NermalTheBunny, Evan-Amos, Bjung, Shaddim, Gary ml0gmwuuc, Khazar2, Gumpboy97, OMPIRE, Cartakes, So-retro-it-hurts and Anonymous: 33

- **Commodore 64 joystick adapters** *Source:* https://en.wikipedia.org/wiki/Commodore_64_joystick_adapters?oldid=687241112 *Contributors:* Kwamikagami, Polluks, Frodet, Electron9, Addbot, Eumolpo, LilHelpa, Stasvasy, Dewritech, Helpful Pixie Bot, BG19bot, Lpanta, Stigsynnes and Anonymous: 5

- **Commodore 64 peripherals** *Source:* https://en.wikipedia.org/wiki/Commodore_64_peripherals?oldid=685484408 *Contributors:* Wernher, Tomlouie, Boffy b, Altenmann, Marcika, Sdfisher, Uzume, Bumm13, Pixel8, Polluks, Nholland~enwiki, Kelly Martin, Firsfron, Woohookitty, Bellhalla, Dah31, Shadyman, BD2412, JIP, Rjwilmsi, Vegaswikian, Dbsanfte, Ian Pitchford, Bgwhite, Digitalme, Crotalus horridus, RussBot, Silverdr, Nikkimaria, Scolaire, SmackBot, Betacommand, Oscarthecat, Chris the speller, Bluebot, Bazonka, Vanessadannenberg, CSWarren, Colonies Chris, Emurphy42, Bogsat, BlackTerror, John, Gang65, Dodgem4s, TastyPoutine, DagErlingSmørgrav, CmdrObot, DevinCook, Astralblue, Neelix, Mblumber, Jaktrip, Marksmithnz, Uspn, Trevyn, Electron9, Monkey Man, Messeng3r, T@nn, Destynova, Stephenchou0722, CommonsDelinker, Nono64, PyroGuy, LordSkitch~enwiki, Multicherry, DeeKay64, Bigdumbdinosaur, Soci64, Sarenne, Somnix, Merman1974, Haseo9999, Fnagaton, Rlendog, Lightmouse, IdreamofJeanie, MikeVitale, Mild Bill Hiccup, Crywalt, Petchboo, Ovalnovel~enwiki, Addbot, Michael480, Calle1970se, Yobot, Fraggle81, AnomieBOT, Locos epraix, Ubcule, Locobot, Babylon4, Chaheel Riens, Kwiki, Jopinder, Plbyrd, Soenke Rahn, Dexter Nextnumber, Stasvasy, Angrytoast, AvicAWB, Sbmeirow, Bomazi, Evan-Amos, Matthiaspaul, Compilation finished successfully, BG19bot, Alphamyon, Mogism, Monkbot, OMPIRE, Mbp303, Wikidalien and Anonymous: 63

- **Commodore 65** *Source:* https://en.wikipedia.org/wiki/Commodore_65?oldid=685010714 *Contributors:* Nixdorf, Wernher, Coldacid, Pixel8, Ylee, Incognito, Devil Master, Polluks, TheIguana, Tabletop, StuartBrady, FlaBot, Mirror Vax, Dinoen, Zotel, Todd Vierling, Silverdr, Kufat, Lipothymia, LeonardoRob0t, AndersL, SmackBot, Rtiainen, RedSpruce, Commander Keane bot, Hmains, Thumperward, Newmanbe, Electron9, Silver Edge, BokicaK, Magioladitis, CommonsDelinker, Sjpeckmore, Sarenne, Alatari, Arjayay, Petchboo, CapnZapp, Addbot, AndersBot, Yobot, AnomieBOT, MastiBot, EmausBot, Frietjes, Compilation finished successfully, BG19bot, Wbsander, Joho345, OMPIRE, Commodoresixtyfive and Anonymous: 29

- **Commodore Datasette** *Source:* https://en.wikipedia.org/wiki/Commodore_Datasette?oldid=659201071 *Contributors:* Liftarn, Smack, Wik, Wernher, Boffy b, Psychonaut, Guanabot, Pixel8, Ylee, Drhex, Jef-Infojef, Dah31, GregorB, Kbdank71, Mirror Vax, Todd Vierling, Where next Columbus?, Rwwww, SmackBot, Arny, Thumperward, Rolypolyman, Pipatron, Kubanczyk, Luminifer, Electron9, Mk*, John a s, Austin-murphy, √2, Destynova, Badn3wz~enwiki, AlleborgoBot, Michael Frind, Snideology, Holothurion, Badmachine, Petchboo, Addbot, Laaknor-Bot, Verbal, Luckas-bot, Yobot, AnomieBOT, Chaheel Riens, Surv1v4l1st, Elpiades, MaGa, Pantergraph, Wbm1058, BattyBot, Dexbot, Sofia Koutsouveli, OMPIRE, SPRDEF and Anonymous: 38

- **Commodore Educator 64** *Source:* https://en.wikipedia.org/wiki/Commodore_Educator_64?oldid=660875119 *Contributors:* Edward, Psycho-naut, Gioto, Magioladitis, SieBot, Zurillion, Petchboo, Addbot, Anders, Luckas-bot, Ubcule, Babylon4, EmausBot, Pantergraph, So-retro-it-hurts and Anonymous: 7

- **Commodore MAX Machine** *Source:* https://en.wikipedia.org/wiki/Commodore_MAX_Machine?oldid=660874874 *Contributors:* Nixdorf, Liftarn, Wwwwolf, Dave Farquhar, Delirium, Slark, Wernher, Psychonaut, Pixel8, Roo72, Incognito, Iamunknown, Gene Nygaard, MIT Trekkie, Angr, FlaBot, Mirror Vax, Random user 39849958, BOT-Superzerocool, SmackBot, Renegadeviking, Electron9, Silver Edge, BokicaK, .ana-condabot, Sarenne, Jamelan, Andy Dingley, LanceBarber, Trivialist, Onomou, Addbot, Lightbot, Luckas-bot, Yobot, Ubcule, EmausBot, Wiki-tanvirBot, Wikiwriter80132, Akazik, IlRoberto88, So-retro-it-hurts and Anonymous: 6

- **Commodore OS** *Source:* https://en.wikipedia.org/wiki/Commodore_OS?oldid=683680748 *Contributors:* Tom-Timmy, Miranche, Luke-Jr, Rastavox, Thumperward, Mdwh, Frap, Racklever, DGG, Rainy666, Addbot, Ronhjones, AnomieBOT, Cavarrone, Xqbot, JohnBlood378, Xorxos, Concernedresident's butler, Aoidh, Gatmaster, ClueBot NG, BattyBot, Hairy LuLu, FoCuSandLeArN, Happyman7, Jrnlist724, Maser-atimanly, Fsandlinux, Zenithworld, Hitesh chawla india, In Correct, Lanctot6000 and Anonymous: 14

- **Commodore REU** *Source:* https://en.wikipedia.org/wiki/Commodore_REU?oldid=668688699 *Contributors:* Dave Farquhar, Gestumblindi, Wernher, Sam Hocevar, Klemen Kocjancic, CanisRufus, Polluks, Spiro Trikaliotis~enwiki, Xenium, Gene Nygaard, Mirror Vax, YurikBot, Cro-talus horridus, Lexi Marie, SmackBot, CmdrObot, GerryJJ, Electron9, Taborgate, Sarenne, Suenarmy, Addbot, AndersBot, Ccureau, Erik9bot, John of Reading, RRabbit42, Kikichugirl, Faizan, Kahtar, Equinox and Anonymous: 12

- **Commodore SX-64** *Source:* https://en.wikipedia.org/wiki/Commodore_SX-64?oldid=660874993 *Contributors:* Eloquence, D, Dave Farquhar, Dysprosia, Zoicon5, Omegatron, Wernher, Morn, Psychonaut, Marcika, Bobblewik, Rich Farmbrough, Pixel8, Ylee, Deathawk, Ae-a, Trevie, GraemeLeggett, Jeremyharmon, Ianthegecko, Mirror Vax, Swtpc6800, Crotalus horridus, Zwobot, Sean Whitton, SmackBot, Skizzik, Bowmanjj, TastyPoutine, Drlegendre, Jesse Viviano, Aldis90, Thijs!bot, Electron9, Darev, BokicaK, Andreas Toth, Mrrrl, VolkovBot, Sarenne, Fnagaton, SkeletorUK, DumZiBoT, Addbot, Lightbot, Luckas-bot, Legobot II, AnomieBOT, ArthurBot, Xqbot, Ubcule, RjwilmsiBot, Az29, ClueBot NG, Helpful Pixie Bot, Wbm1058, ChrisGualtieri, OMPIRE, So-retro-it-hurts and Anonymous: 38

- **Compunet** *Source:* https://en.wikipedia.org/wiki/Compunet?oldid=664944626 *Contributors:* Liftarn, Nv8200pa, Loganberry, Mairi, Mirror Vax, Str1977, ONEder Boy, S33k3r, JLaTondre, Mewcenary, Mhardcastle, SmackBot, Maelwys, Colonies Chris, DeFacto, Nick Green, Wag-gers, DabMachine, CmdrObot, Cydebot, Bencherlite, Wikiisawesome, Umayxa3, Malcolmxl5, Dvdvnr, Czarkoff, Eeekster, Petchboo, Addbot, AkhtaBot, Luckas-bot, Julle, Chaheel Riens, FrescoBot, PartTimeGnome, Alexblewitt, Mjb1962 and Anonymous: 6

- **Creative Micro Designs** *Source:* https://en.wikipedia.org/wiki/Creative_Micro_Designs?oldid=664530667 *Contributors:* ZoeB, Wernher, Hadal, Coldacid, Anirvan, MakeRocketGoNow, Polluks, RJFJR, Nholland~enwiki, Mirror Vax, Jaraalbe, Crotalus horridus, Epolk, Nil Einne, Hmains, Thumperward, DMS, JonHarder, TPO-bot, GerryJJ, Cydebot, Saruwine, Addbot, Download, Lightbot, Yobot, AnomieBOT, DrilBot, Irkirkirk, Jonpatterns, Traymnd, ShaunBebbers and Anonymous: 15

- **DolphinDOS** *Source:* https://en.wikipedia.org/wiki/DolphinDOS?oldid=580861341 *Contributors:* Bearcat, JIP, Silverdr, GoodDay, Katharineamy, Ironholds and Dexter Nextnumber

- **Epyx Fast Load** *Source:* https://en.wikipedia.org/wiki/Epyx_Fast_Load?oldid=641789395 *Contributors:* Frecklefoot, Dave Farquhar, Wernher, Ylee, Alai, Mirror Vax, Crotalus horridus, RussBot, SmackBot, Sgt Pinback, J00tel, Sjhoran, NorthernThunder, Gwern, Knyte6426, Mpeylo, Addbot, Lightbot, RenamedUser01302013, Sbmeirow and Anonymous: 8

- **GeoRAM** *Source:* https://en.wikipedia.org/wiki/GeoRAM?oldid=672243746 *Contributors:* Bryan Derksen, Wernher, Bumm13, FlaBot, Mirror Vax, SmackBot, Thijs!bot, Addbot, Erik9bot, Amina.alobaid, So-retro-it-hurts and Anonymous: 3

- **Human Engineered Software** *Source:* https://en.wikipedia.org/wiki/Human_Engineered_Software?oldid=660035220 *Contributors:* Wernher, Inkling, Ylee, Longhair, FlaBot, Mirror Vax, Krótki, Vanka5, Crystallina, SmackBot, Hmains, Mairibot, Emurphy42, Jinnai, Jeffmalmberg, Cydebot, Alaibot, BetacommandBot, Visik, Magioladitis, Cander0000, Gwern, Xgmx, Falcon8765, TJ Moose, Stepheng3, Addbot, Lightbot, MuZemike, Erik9bot, I dream of horses, KarlsenBot, IPhoneGamer32, OMPIRE and Anonymous: 8

- **IDE64** *Source:* https://en.wikipedia.org/wiki/IDE64?oldid=556849415 *Contributors:* Damian Yerrick, Nufy8, Closeapple, Polluks, FlaBot, Gaius Cornelius, SmackBot, Thumperward, Electron9, Bigdumbdinosaur, Soci64, Signalhead, Saber girl08, Addbot, LilHelpa, DrilBot, Bat-tyBot and Anonymous: 10

- **Individual Computers** *Source:* https://en.wikipedia.org/wiki/Individual_Computers?oldid=635301823 *Contributors:* Wernher, Sysy, Polluks, Cavrdg, MagerValp, Alarob, Crystallina, Mairibot, Electron9, DeeKay64, Caltas, Lightbot, Fox Wilson, Twistedtrick and Anonymous: 9

- **Indus GT** *Source:* https://en.wikipedia.org/wiki/Indus_GT?oldid=653015615 *Contributors:* Polluks, Nick Number, The Elves Of Dunsimore, Ubcule, FrescoBot, Pro-7800, Compilation finished successfully, OMPIRE and Anonymous: 3

- **ISEPIC** *Source:* https://en.wikipedia.org/wiki/ISEPIC?oldid=676920729 *Contributors:* Ylee, JIP, Thumperward, Electron9, Mild Bill Hiccup, Yobot and ChrisGualtieri

- **KERNAL** *Source:* https://en.wikipedia.org/wiki/KERNAL?oldid=681890064 *Contributors:* Nixdorf, Silvonen, Furrykef, Grendelkhan, Head, Wernher, Aenar, Tobias Bergemann, Bumm13, Andreas Kaufmann, Abdull, Zondor, Twenex, Pixel8, Ylee, MagerValp, MarkusHagenlocher, JIP, Mirror Vax, PhilipO, Cedar101, Reedy, Xaosflux, Thumperward, Miquonranger03, Random Kingdom, Cydebot, Ntsimp, Mblumber, Gogo Dodo, Alaibot, Thijs!bot, Bongwarrior, Nono64, Bigdumbdinosaur, VolkovBot, Sarenne, Allanth, Jmath666, Caltas, ClueBot, Dpmuk, XLinkBot, MystBot, Addbot, Mortense, Lightbot, Fraggle81, WorkingBeaver, Quebec99, PigFlu Oink, Fox Wilson, Dexter Nextnumber, De-sihacker, Midas02, Owenmann, ClueBot NG, Pantergraph, Jens8742, Daroooo, AsusStealth, Ricochet Jones, Mogism, Param Mudgal, Kkm-mdd9069, Jayarajr007, SPRDEF and Anonymous: 43

- **Lt. Kernal** *Source:* https://en.wikipedia.org/wiki/Lt._Kernal?oldid=649711790 *Contributors:* Pnm, Psychonaut, Coldacid, TheParanoidOne, BD2412, JIP, Angusmclellan, Robertvan1, Rwwww, SmackBot, Bigdumbdinosaur, Lightbot, Smuckola, OMPIRE and Anonymous: 1

- **Magnum Light Phaser** *Source:* https://en.wikipedia.org/wiki/Magnum_Light_Phaser?oldid=544102623 *Contributors:* Boffy b, Polluks, TheParanoidOne, StuartBrady, Larsinio, Whobot, David Wahler, Mairibot, Bluebot, Abhisara, Alaibot, Merman1974, Addbot, Lightbot, Luckas-bot, YFdyh-bot and Anonymous: 3

- **MMC64** *Source:* https://en.wikipedia.org/wiki/MMC64?oldid=622578544 *Contributors:* Uzume, Jeff3000, Vossanova, Todd Vierling, Gaius Cornelius, SmackBot, Cmh, Hubba, Electron9, Rlbberlin, MarshBot, Steevven1, CommonsDelinker, DeeKay64, Signalhead, Addbot, Hanzu777, Bomazi, Helpful Pixie Bot, Alphamyon, Kephir and Anonymous: 11

- **MOS Technology 6510** *Source:* https://en.wikipedia.org/wiki/MOS_Technology_6510?oldid=670944121 *Contributors:* Stephen Gilbert, Matusz, Nixdorf, Wernher, Pixel8, Overmann, Wtshymanski, Ae-a, Marudubshinki, FlaBot, Mirror Vax, YurikBot, Crotalus horridus, Malcolma, Klaws, AndersL, SmackBot, Rtiainen, CSWarren, Tlusťa, Gang65, Sjf, Grasshoppa, Ronaldvd, Thijs!bot, Al Lemos, JAnDbot, Nono64, Bigdumbdinosaur, Fnagaton, Wdwd, Niceguyedc, Alexbot, InternetMeme, Jklowden, Legobot, Yobot, ArthurBot, ZéroBot, Helpful Pixie Bot, Taibah U, BG19bot, Hmainsbot1, TortoiseWrath, Comp.arch and Anonymous: 24

- **MOS Technology 8502** *Source:* https://en.wikipedia.org/wiki/MOS_Technology_8502?oldid=653374609 *Contributors:* Wernher, Polluks, Morkork, Mirror Vax, Crotalus horridus, Malcolma, Dinjiin, Gang65, Ronaldvd, Electron9, RareEntity, SieBot, Raffzahn, Shape84, Alexbot, Addbot, FrescoBot, ZéroBot, ChuispastonBot, Matthiaspaul and Anonymous: 8

- **MOS Technology 8563** *Source:* https://en.wikipedia.org/wiki/MOS_Technology_8563?oldid=669851005 *Contributors:* Maury Markowitz, Dave Farquhar, Wernher, Tjansen, Pixel8, Polluks, Mirror Vax, Russavia, Crotalus horridus, CSWarren, Radagast83, Ronaldvd, Thijs!bot, Alphachimpbot, Vanessaezekowitz, Nono64, Bigdumbdinosaur, Sarenne, Shape84, Addbot, Ccureau, Xqbot, CoolingGibbon, Hmainsbot1 and Anonymous: 14

- **MOS Technology 8568** *Source:* https://en.wikipedia.org/wiki/MOS_Technology_8568?oldid=669855232 *Contributors:* Wernher, Altenmann, Centrx, Polluks, Wtshymanski, FlaBot, Mirror Vax, Crotalus horridus, Musicomputer, Radagast83, FairuseBot, Ronaldvd, Electron9, Alphachimpbot, Xeno, R'n'B, Bigdumbdinosaur, Sarenne, Lightmouse, Addbot, AnomieBOT, CoolingGibbon, RedBot, Updatehelper and Anonymous: 4

- **MOS Technology CIA** *Source:* https://en.wikipedia.org/wiki/MOS_Technology_CIA?oldid=652481722 *Contributors:* Uzume, Hugh Mason, Polluks, Hooperbloob, BD2412, Mirror Vax, YurikBot, Crotalus horridus, RussBot, Silverdr, Grafen, Rtiainen, Ronaldvd, Thijs!bot, Electron9, Marko75, Daibot~enwiki, Nono64, Bigdumbdinosaur, Spinningspark, WikHead, Addbot, Ettrig, Luckas-bot, JWBE, John of Reading, 220 of Borg and Anonymous: 10

- **MOS Technology SID** *Source:* https://en.wikipedia.org/wiki/MOS_Technology_SID?oldid=683998187 *Contributors:* Robert Merkel, Taw, Christian List, D, Lezek, Lexor, DopefishJustin, Mahjongg, Nixdorf, Liftarn, Wwwwolf, ZoeB, Notheruser, Tedius Zanarukando, Malcohol, Themaxx, Slark, WhisperToMe, Furrykef, Wernher, Jeeves, Twang, Robbot, RedWolf, Dehumanizer, Kevin Saff, Honta, Ds13, Spm, Edcolins, Chowbok, LucasVB, Mamizou, Qdr, Perey, Sysy, Pixel8, Roo72, Ylee, RoyBoy, Dgpop, JRM, Polluks, Trevj, PJ, Ashley Pomeroy, Wtshymanski, Cburnett, Derbeth, Reaverdrop, Voxadam, Forderud, Jef-Infojef, Ae-a, Al E., Bluemoose, D14BL0, Graham87, Rjwilmsi, Captain Disdain, Mirror Vax, SchuminWeb, Ysangkok, Viznut, Liontamer, Flashmorbid, Bgwhite, NSR, YurikBot, Crotalus horridus, RmM, Zafiroblue05, FromWithin, CapitalLetterBeginning, NotInventedHere, Vampyrium, 2fort5r, KnightRider~enwiki, SmackBot, Eskimbot, Fuzzform, Frap, Paulie68000, Deepred6502, PXE-M0F, JMax555, Danmoore, Wickethewok, Anescient, Tawkerbot2, ChrisCork, Ronaldvd, ShelfSkewed, WeggeBot, Mccalli, Cydebot, Silvertouch57, Lord Nightmare, Al Lemos, Electron9, X201, AntiVandalBot, Darkuni, Infindebula, Damian1983, Kaini, Daibot~enwiki, Stolsvik, Conquerist, Scoutski, Vanessaezekowitz, Rettetast, Verdatum, Wikip rhyre, Dispenser, Gigantic Killerdong, DeeKay64, BOTones, JJGD, Jcea, Leopold B. Stotch, Jalwikip, Billinghurst, Spinningspark, Nunucello, Rlendog, Lightmouse, Peepo uk, Martarius, EoGuy, Niceguyedc, Suenarmy, Rhododendrites, Pinkevin, Theinvisibleworm, InternetMeme, XLinkBot, Duncan, Imanjl, Addbot, Yobot, Pcap, Chordian, AnomieBOT, MauritsBot, Xqbot, Cybjit, C64glen, SimonInns, RedBot, Jopinder, Johnnylocust, RjwilmsiBot, EmausBot, Alfons van Vorden, ZéroBot, Plm-pro, A930913, Kent Sullivan, ClueBot NG, DieSwartzPunkt, Cntras, Park Flier, Molotovsystem, Yoyofr, AndroSID, BattyBot, Thomas.hori, Scott Samwell, BaseCochise, Monkbot and Anonymous: 207

- **MOS Technology VIC-II** *Source:* https://en.wikipedia.org/wiki/MOS_Technology_VIC-II?oldid=679339411 *Contributors:* Mahjongg, Nixdorf, Liftarn, Wwwwolf, Wernher, Coldacid, Pixel8, Roo72, JRM, Polluks, DaveGorman, GregorB, Teknic, Mirror Vax, Viznut, YurikBot, Crotalus horridus, DragonHawk, Richardcavell, That Guy, From That Show!, Krótki, SmackBot, Fuhghettaboutit, Pemu, Howdoesthiswo, Gang65, Ronaldvd, J Milburn, Mblumber, LarryMColeman, Lugnuts, JLD, Electron9, .anacondabot, R'n'B, DeeKay64, Chtaube~enwiki, Jalwikip, Sarenne, McM.bot, AlleborgoBot, Abc64, SchreiberBike, Addbot, E5frog, Lightbot, Zorrobot, Luckas-bot, Yobot, Hanzu777, AnomieBOT, Quebec99, CoolingGibbon, Glider87, Dcirovic, Wcherowi, Kevonni, BG19bot, 4throck, Digital Brains, BattyBot, Khazar2, Monkbot and Anonymous: 47

- **PETSCII** *Source:* https://en.wikipedia.org/wiki/PETSCII?oldid=666286871 *Contributors:* Mjb, Mahjongg, Nixdorf, Wernher, Psychonaut, DavidCary, Mboverload, Bumm13, Abdull, PhennPhawcks, Rich Farmbrough, Lovelac7, Jaberwocky6669, Shlomital, Wtshymanski, Apokrif, Gudeldar, FlaBot, Mirror Vax, Random user 39849958, Zwobot, Calcwatch, Krótki, SmackBot, Radak, Loadmaster, TastyPoutine, Seek100, DevinCook, Nczempin, Thijs!bot, Gioto, Magioladitis, R'n'B, Kernal 7.1, Austriacus, Ogre lawless, DumZiBoT, Addbot, Yobot, Coroboy, Uhmgawa, LawBot, Nevin.williams, Angrytoast, Akazik, Pantergraph, Matthiaspaul, BG19bot, Digital Brains, Synthetoonz, MartUK2012 and Anonymous: 33

- **PlayNET** *Source:* https://en.wikipedia.org/wiki/PlayNET?oldid=667970634 *Contributors:* BRG, Selket, Ylee, Enric Naval, Padawer, Jesup, RocketMaster, AKMask, J00tel, ShelfSkewed, Cydebot, DuncanHill, Appraiser, BobF4, Drunkenmonkey, HowardSG, Android Mouse Bot 3, XLinkBot, ENGINR, Yobot, AnomieBOT, HowardSG16 and Anonymous: 13

- **Punter (protocol)** *Source:* https://en.wikipedia.org/wiki/Punter_(protocol)?oldid=685838470 *Contributors:* Maury Markowitz, PBS, Psychonaut, Rpresser, Cwolfsheep, Mirror Vax, Bhny, SmackBot, Emurphy42, ClueBot, Qwetzalquoatal, HexaChord, Lightbot, AnomieBOT, Kristen Eriksen, Palosirkka and Anonymous: 12

- **Quantum Link** *Source:* https://en.wikipedia.org/wiki/Quantum_Link?oldid=672509619 *Contributors:* Dwheeler, Edward, DavidWBrooks, Arteitle, Crissov, Zoicon5, Joy, Boffy b, DocWatson42, Willhester, Murple, Kate, CanisRufus, CyberSkull, Ricky81682, Fritz Saalfeld, Strongbow, Sburke, Jquarry, Mirror Vax, Zotel, Jesup, SmackBot, ShelfSkewed, Cydebot, Warhorus, JamesLucas, Vanessaezekowitz, RJaguar3, Musicandnintendo, Lightmouse, ImageRemovalBot, Hx823, Petchboo, Inemanja, Babylon4, Krinkle, Sbmeirow, Ramaksoud2000 and Anonymous: 35

- **Scene World Magazine** *Source:* https://en.wikipedia.org/wiki/Scene_World_Magazine?oldid=687329151 *Contributors:* Derision667, XLinkBot, Yobot, WikiDan61, Cnwilliams, BG19bot, Joseph2302, Dummyworker123 and Anonymous: 6

- **Stack Light Rifle** *Source:* https://en.wikipedia.org/wiki/Stack_Light_Rifle?oldid=644677219 *Contributors:* Rich Farmbrough, Pak21, Polluks, Jtalledo, Welsh, Larsinio, Whobot, JLaTondre, SmackBot, Bluebot, OrphanBot, Remember the dot, WOSlinker, Yobot and Anonymous: 5

- **Super 1750 Clone** *Source:* https://en.wikipedia.org/wiki/Super_1750_Clone?oldid=561805811 *Contributors:* Delirium, Wernher, SmackBot, Thijs!bot, Electron9, Addbot, Uzma Gamal, Amina.alobaid and Anonymous: 6

- **Super Expander 64** *Source:* https://en.wikipedia.org/wiki/Super_Expander_64?oldid=464498991 *Contributors:* Polluks, Michael Drüing, Marudubshinki, Wizardman, Euchiasmus, CmdrObot, The Thing That Should Not Be, John of Reading and Anonymous: 5

- **SuperCPU** *Source:* https://en.wikipedia.org/wiki/SuperCPU?oldid=670236027 *Contributors:* Polluks, Woohookitty, Tomconte, FlaBot, Nikkimaria, SmackBot, A876, Electron9, Omega-Q, Addbot, Jonpatterns, Frank Ulysses Cofferd, Wbm1058, ChrisGualtieri, Jodosma, Niculinux, OMPIRE and Anonymous: 5

- **The Final Cartridge III** *Source:* https://en.wikipedia.org/wiki/The_Final_Cartridge_III?oldid=682260097 *Contributors:* Liftarn, Wwwwolf, Wernher, Rossumcapek, Jawed, Craigy144, Dowcet, Kelly Martin, YurikBot, Thumperward, Electron9, Magioladitis, GrahamHardy, Addbot, AnomieBOT, EmausBot, Wbm1058, Fmb64, Cryofreezer and Anonymous: 8

- **The Judges (demogroup)** *Source:* https://en.wikipedia.org/wiki/The_Judges_(demogroup)?oldid=643086311 *Contributors:* Liftarn, Frodet, Vossanova, JIP, Viznut, Aeternus, Cydebot, Phil van zyl, Spartaz, GrahamHardy, ClueBot, Yobot, Minus3 and Anonymous: 3

- **Trilogic Expert Cartridge** *Source:* https://en.wikipedia.org/wiki/Trilogic_Expert_Cartridge?oldid=514628475 *Contributors:* Nixdorf, Thumperward, Electron9 and EoGuy

- **Ventilator 202** *Source:* https://en.wikipedia.org/wiki/Ventilator_202?oldid=667020181 *Contributors:* Bearcat, Picapica, Ricky81682, Iothiania, Open2universe, Ajdebre, SmackBot, Zvonko, Aleksandar Šušnjar, RomanSpa, After Midnight, Sjar, Niceguyedc, Milosppf, Dawynn, Lightbot, Yobot, FrescoBot, Tomtommusic, John of Reading, ChrisGualtieri and Anonymous: 6

- **Xetec** *Source:* https://en.wikipedia.org/wiki/Xetec?oldid=622317512 *Contributors:* Uzume, D6, SmackBot, Kjaergaard, Sgt Pinback, Deflective, Marko75, Emeraude, PyroGuy, Strange deja vu, Bigdumbdinosaur, One Elephant went out to play..., Smuckola, Sbmeirow, BattyBot and Anonymous: 3

3.2 Images

- **File:1541_ultimate_plus.jpg** *Source:* https://upload.wikimedia.org/wikipedia/commons/d/d0/1541_ultimate_plus.jpg *License:* CC BY-SA 3.0 *Contributors:* Own work *Original artist:* Hedning

- **File:6510_CPU_Pinout.svg** *Source:* https://upload.wikimedia.org/wikipedia/commons/e/e0/6510_CPU_Pinout.svg *License:* CC-BY-SA-3.0 *Contributors:* Transferred from en.wikipedia to Commons. *Original artist:* Crotalus horridus at English Wikipedia

- **File:6526_CIA_Pinout.svg** *Source:* https://upload.wikimedia.org/wikipedia/commons/7/72/6526_CIA_Pinout.svg *License:* CC-BY-SA-3.0 *Contributors:* Own work *Original artist:* Crotalus_horridus (talk) (Uploads)

- **File:6581R4_in_cermaic_DIP_manufactured_in_week_11_of_1986.jpg** *Source:* https://upload.wikimedia.org/wikipedia/commons/c/cc/6581R4_CDIP_1186.jpg *License:* CC BY-SA 3.0 *Contributors:* Own work *Original artist:* androSID

- **File:6581_in_ceramic_DIP.jpg** *Source:* https://upload.wikimedia.org/wikipedia/commons/0/05/6581_in_ceramic_DIP.jpg *License:* CC BY-SA 3.0 *Contributors:* Own work *Original artist:* androSID

- **File:6582A.jpg** *Source:* https://upload.wikimedia.org/wikipedia/commons/0/0b/6582A.jpg *License:* CC BY-SA 3.0 *Contributors:* Own work *Original artist:* androSID

- **File:8580R5_USA.jpg** *Source:* https://upload.wikimedia.org/wikipedia/commons/f/f4/8580R5_USA.jpg *License:* CC BY-SA 3.0 *Contributors:* Own work *Original artist:* androSID

- **File:Action_Replay_Amiga500.jpg** *Source:* https://upload.wikimedia.org/wikipedia/commons/8/81/Action_Replay_Amiga500.jpg *License:* CC BY-SA 2.5 *Contributors:* Own work *Original artist:* Afrank99

- **File:Action_Replay_C64.jpg** *Source:* https://upload.wikimedia.org/wikipedia/commons/9/91/Action_Replay_C64.jpg *License:* Public domain *Contributors:* Own work *Original artist:* joho345

- **File:Action_replay_ds.JPG** *Source:* https://upload.wikimedia.org/wikipedia/commons/e/ea/Action_replay_ds.JPG *License:* CC BY-SA 2.5 *Contributors:* Own work *Original artist:* Laura Ohrndorf

- **File:Ambox_important.svg** *Source:* https://upload.wikimedia.org/wikipedia/commons/b/b4/Ambox_important.svg *License:* Public domain *Contributors:* Own work, based off of Image:Ambox scales.svg *Original artist:* Dsmurat (talk · contribs)

- **File:Ambox_rewrite.svg** *Source:* https://upload.wikimedia.org/wikipedia/commons/1/1c/Ambox_rewrite.svg *License:* Public domain *Contributors:* self-made in Inkscape *Original artist:* penubag

- **File:C-One_motherboard.jpg** *Source:* https://upload.wikimedia.org/wikipedia/commons/4/4b/C-One_motherboard.jpg *License:* CC BY 2.0 *Contributors:* originally posted to **Flickr** as what's where on the C-One motherboard *Original artist:* Blake Patterson

- **File:C2n_waveform.png** *Source:* https://upload.wikimedia.org/wikipedia/commons/c/c3/C2n_waveform.png *License:* CC BY-SA 3.0 *Contributors:* http://en.wikipedia.org/wiki/File:C2n_waveform.png *Original artist:* Electron9 on english Wikipedia

- **File:C64-IMG_5372.jpg** *Source:* https://upload.wikimedia.org/wikipedia/commons/9/90/C64-IMG_5372.jpg *License:* Public domain *Contributors:* Own work *Original artist:* Kausalkette

- **File:C64Cmotherboard.jpg** *Source:* https://upload.wikimedia.org/wikipedia/commons/9/93/C64Cmotherboard.jpg *License:* CC-BY-SA-3.0 *Contributors:* Bill Bertram *Original artist:* Bill Bertram

- **File:C64GS-Console-Set.jpg** *Source:* https://upload.wikimedia.org/wikipedia/commons/f/fd/C64GS-Console-Set.jpg *License:* CC BY-SA 3.0 *Contributors:* Own work *Original artist:* Evan-Amos

- **File:C64_Petscii_Charts.png** *Source:* https://upload.wikimedia.org/wikipedia/commons/c/c4/C64_Petscii_Charts.png *License:* Public domain *Contributors:* ? *Original artist:* ?

- **File:C64_startup_animiert.gif** *Source:* https://upload.wikimedia.org/wikipedia/commons/4/48/C64_startup_animiert.gif *License:* Public domain *Contributors:* ? *Original artist:* ?

- **File:C64c_system.jpg** *Source:* https://upload.wikimedia.org/wikipedia/commons/8/84/C64c_system.jpg *License:* CC BY-SA 2.5 *Contributors:* Own work *Original artist:* Bill Bertram

- **File:C64motherboard.jpg** *Source:* https://upload.wikimedia.org/wikipedia/commons/8/86/C64motherboard.jpg *License:* CC BY 2.0 *Contributors:* Bill Bertram *Original artist:* Bill Bertram

- **File:C64plus-IEEE488.JPG** *Source:* https://upload.wikimedia.org/wikipedia/commons/8/81/C64plus-IEEE488.JPG *License:* CC BY-SA 3.0 *Contributors:* Transferred from en.wikipedia to Commons by Undead_warrior. *Original artist:* Suenarmy at English Wikipedia

- **File:C65-open.jpg** *Source:* https://upload.wikimedia.org/wikipedia/commons/1/13/C65-open.jpg *License:* CC BY 2.5 *Contributors:* ? *Original artist:* ?

- **File:C65.png** *Source:* https://upload.wikimedia.org/wikipedia/commons/c/c9/C65.png *License:* Public domain *Contributors:* Transferred from de.wikipedia to Commons. *Original artist:* The original uploader was Tom Knox at German Wikipedia

- **File:C65cpu.jpg** *Source:* https://upload.wikimedia.org/wikipedia/commons/1/1a/C65cpu.jpg *License:* CC-BY-SA-3.0 *Contributors:* Own work *Original artist:* Machine Later versions were uploaded by Chaddy.

- **File:C65elmer.jpg** *Source:* https://upload.wikimedia.org/wikipedia/commons/6/64/C65elmer.jpg *License:* CC-BY-SA-3.0 *Contributors:* Own work *Original artist:* Machine Later versions were uploaded by Chaddy.

- **File:C65mainboard.jpg** *Source:* https://upload.wikimedia.org/wikipedia/commons/d/d1/C65mainboard.jpg *License:* CC-BY-SA-3.0 *Contributors:* Own work *Original artist:* Machine

- **File:C65offen.jpg** *Source:* https://upload.wikimedia.org/wikipedia/commons/f/f9/C65offen.jpg *License:* CC-BY-SA-3.0 *Contributors:* Own work *Original artist:* Machine

- **File:C65u22.jpg** *Source:* https://upload.wikimedia.org/wikipedia/commons/5/50/C65u22.jpg *License:* CC-BY-SA-3.0 *Contributors:* Own work *Original artist:* Benutzer:Machine.

- **File:C65viciii.jpg** *Source:* https://upload.wikimedia.org/wikipedia/commons/7/76/C65viciii.jpg *License:* CC-BY-SA-3.0 *Contributors:* Own work *Original artist:* Machine Later versions were uploaded by Chaddy.

- **File:CBM64Cartridges.JPG** *Source:* https://upload.wikimedia.org/wikipedia/commons/a/ad/CBM64Cartridges.JPG *License:* Public domain *Contributors:* Own work *Original artist:* ML5

- **File:CBM_Logo.svg** *Source:* https://upload.wikimedia.org/wikipedia/commons/5/5c/CBM_Logo.svg *License:* Public domain *Contributors:* This file has been **extracted** from another file: Commodore logo.svg.
  Original artist: Commodore Int. (Commodore logo and typeface created in inkscape 0.44 by Pixel8.)

- **File:CMDLogo.gif** *Source:* https://upload.wikimedia.org/wikipedia/en/7/74/CMDLogo.gif *License:* Fair use *Contributors:* http://www.c64-wiki.de/images/7/74/CMDLogo.gif *Original artist:* ?

- **File:COMPUTAPIX+.jpg** *Source:* https://upload.wikimedia.org/wikipedia/en/7/79/COMPUTAPIX%2B.jpg *License:* GFDL *Contributors:* Own work
  Original artist:
  Suenarmy (talk) (Uploads)

- **File:CSG_6582A.jpg** *Source:* https://upload.wikimedia.org/wikipedia/commons/5/56/CSG_6582A.jpg *License:* CC BY-SA 3.0 *Contributors:* Own work *Original artist:* AndroSID

- **File:Cbmcharset-modes.png** *Source:* https://upload.wikimedia.org/wikipedia/en/f/f5/Cbmcharset-modes.png *License:* Cc-by-sa-3.0 *Contributors:* ? *Original artist:* ?

- **File:Commodore-64-Back.jpg** *Source:* https://upload.wikimedia.org/wikipedia/commons/c/c9/Commodore-64-Back.jpg *License:* Public domain *Contributors:* Own work *Original artist:* Evan-Amos

- **File:Commodore-64-Computer.jpg** *Source:* https://upload.wikimedia.org/wikipedia/commons/f/f4/Commodore-64-Computer.jpg *License:* Public domain *Contributors:* Own work *Original artist:* Evan-Amos

- **File:Commodore-64-Computer.png** *Source:* https://upload.wikimedia.org/wikipedia/commons/3/34/Commodore-64-Computer.png *License:* Public domain *Contributors:* Own work *Original artist:* Evan-Amos

- **File:Simons_Basic_Splash_Screen.gif** *Source:* https://upload.wikimedia.org/wikipedia/en/9/96/Simons_Basic_Splash_Screen.gif *License:* PD *Contributors:* ? *Original artist:* ?

- **File:Sinclair_magnum_light_phaser.jpg** *Source:* https://upload.wikimedia.org/wikipedia/commons/8/81/Sinclair_magnum_light_phaser. jpg *License:* CC BY 2.5 *Contributors:* Own work *Original artist:* Boffy b

- **File:SpeedScript_128_in_action.png** *Source:* https://upload.wikimedia.org/wikipedia/commons/f/fe/SpeedScript_128_in_action.png *License:* Public domain *Contributors:* ? *Original artist:* ?

- **File:Steve_Punter.jpg** *Source:* https://upload.wikimedia.org/wikipedia/commons/e/e4/Steve_Punter.jpg *License:* CC BY-SA 3.0 *Contributors:* BBS: The Documentary (2005), specifically from this file *Original artist:* Jason Scott

- **File:Supratechnic_Demo.gif** *Source:* https://upload.wikimedia.org/wikipedia/en/5/51/Supratechnic_Demo.gif *License:* ? *Contributors:* ? *Original artist:* ?

- **File:Sx-64_build_crop.jpg** *Source:* https://upload.wikimedia.org/wikipedia/commons/7/78/Sx-64_build_crop.jpg *License:* CC-BY-SA-3.0 *Contributors:* This file was derived from Sx-64 build.jpg: <a href='//commons.wikimedia.org/wiki/File:Sx-64_build.jpg' class='image'><img alt='Sx-64 build.jpg' src='https://upload.wikimedia.org/wikipedia/commons/thumb/5/5a/Sx-64_build.jpg/130px-Sx-64_build.jpg' width='130' height='98' srcset='https://upload.wikimedia.org/wikipedia/commons/thumb/5/5a/Sx-64_build.jpg/195px-Sx-64_build.jpg 1.5x, https://upload.wikimedia.org/wikipedia/commons/thumb/5/5a/Sx-64_build.jpg/260px-Sx-64_build.jpg 2x' data-file-width='800' data-file-height='600' /></a>
  Original artist:

- **Sx-64_build.jpg: Themaverick**

- **File:Technofor-IEEE488.JPG** *Source:* https://upload.wikimedia.org/wikipedia/en/4/48/Technofor-IEEE488.JPG *License:* GFDL *Contributors:* ? *Original artist:* ?

- **File:Text_document_with_red_question_mark.svg** *Source:* https://upload.wikimedia.org/wikipedia/commons/a/a4/Text_document_with_ red_question_mark.svg *License:* Public domain *Contributors:* Created by bdesham with Inkscape; based upon Text-x-generic.svg from the Tango project. *Original artist:* Benjamin D. Esham (bdesham)

- **File:TheFinalCartridge3.png** *Source:* https://upload.wikimedia.org/wikipedia/en/f/f1/TheFinalCartridge3.png *License:* ? *Contributors:* ? *Original artist:* ?

- **File:Tux.svg** *Source:* https://upload.wikimedia.org/wikipedia/commons/3/35/Tux.svg *License:* Attribution *Contributors:* [1] *Original artist:* Larry Ewing, Simon Budig, Anja Gerwinski

- **File:Ultra_Hi-Res_Cube_Demo.gif** *Source:* https://upload.wikimedia.org/wikipedia/en/c/c0/Ultra_Hi-Res_Cube_Demo.gif *License:* ? *Contributors:* ? *Original artist:* ?

- **File:Unbalanced_scales.svg** *Source:* https://upload.wikimedia.org/wikipedia/commons/f/fe/Unbalanced_scales.svg *License:* Public domain *Contributors:* ? *Original artist:* ?

- **File:VC40_Cart.jpg** *Source:* https://upload.wikimedia.org/wikipedia/commons/7/79/VC40_Cart.jpg *License:* Public domain *Contributors:* Transferred from en.wikipedia to Commons by Logan using CommonsHelper. *Original artist:* Suenarmy at English Wikipedia

- **File:VIC-II.svg** *Source:* https://upload.wikimedia.org/wikipedia/commons/d/da/VIC-II.svg *License:* CC BY 2.5 *Contributors:* Bill Bertram *Original artist:* Bill Bertram

- **File:VIC-II_color_map.svg** *Source:* https://upload.wikimedia.org/wikipedia/commons/5/52/VIC-II_color_map.svg *License:* Public domain *Contributors:* http://en.wikipedia.org/wiki/Image:VIC-II_color_map.svg *Original artist:*
  Crotalus horridus at the English language Wikipedia, the copyright holder of this work, hereby publishes it under the following license:

- **File:Wiki_letter_w_cropped.svg** *Source:* https://upload.wikimedia.org/wikipedia/commons/1/1c/Wiki_letter_w_cropped.svg *License:* CC-BY-SA-3.0 *Contributors:*

- Wiki_letter_w.svg *Original artist:* Wiki_letter_w.svg: Jarkko Piiroinen

- **File:Winchester-Festplatte.jpg** *Source:* https://upload.wikimedia.org/wikipedia/commons/0/0e/Winchester-Festplatte.jpg *License:* Public domain *Contributors:* Transferred from de.wikipedia to Commons.
  Original artist: Winhistory at German Wikipedia

3.3 Content license

www.ingramcontent.com/pod-product-compliance
Lightning Source LLC
Chambersburg PA
CBHW080812180526
45168CB00006B/2415